ROCKET GIRL

A Stage Play By

GEORGE D MORGAN

HS 1

"In the 1950's, Mary Morgan single handedly
saved America's space program, and
nobody knows about it but a handful of old men."

--Walter Unterberg, aerospace engineer (ret.)

**To arrange for the performance rights for
Rocket Girl, please go to:
www.georgedmorgan.com/services**

Rocket Girl was presented at the Ramo Auditorium
Theatre of the California Institute of Technology in
Pasadena, California, on November 7, 2008 by TACIT.
It was directed by Brian Brophy.

CARLEY MORGAN	Meg Rosenburg
BARBARA	Hui Ying Wen
BILL WEBER	Cliff Chang
JOE FRIEDMAN	Jon Napolitano
IRVING KANAREK	Kevin Welch
DON JENKINS	Garrett Lewis
TOM MEYERS	David Seal
MARY SHERMAN	Christina Kondos
GENERAL MEDARIS	Doug Smith
MISS BIDDLE	Miranda Stewart
COLONEL WILKINS	Todd Brun
MICHAEL SHERMAN	Steve Collins
BETTY MANNING	Faith Shuker-Haines
RICHARD MORGAN	Jim Carnesky
MRS. WALKER	Cara King
DOCTOR	Hui Ying Wen
DUTCH KINDELBERGER	Craig Peterson
PAUL MORSKY	Doug Smith
FATHER MACKEY	Zachary Abbott
DAVID SHELBY	Zachary Abbott
PARK RANGER	Miranda Stewart

Directed By BRIAN BROPHY

Written By GEORGE D MORGAN

Produced By TACIT - Theatre Arts of the
California Institute of Technology
Pasadena, California

CHARACTERS - In order of appearance

HS 3

CARLEY MORGAN	20's, college student
BARBARA	Over 30, obit department manager
BILL WEBER	Early 20's, aerospace engineer
JOE FRIEDMAN	Late 20's, aerospace engineer
IRVING KANAREK[1]	Early 30's, aerospace engineer
DON JENKINS	Early 20's, aerospace engineer
TOM MEYERS	40's, aerospace dept. manager
MARY SHERMAN	Late 20's, engineering Analyst
GENERAL MEDARIS[2]	50's, 3-star Army general
MISS BIDDLE	General's secretary
COLONEL WILKINS	40's, Army colonel
MICHAEL SHERMAN	50's, Mary's father
BETTY MANNING	Any adult age, social worker
RICHARD MORGAN	Early 20's, aerospace engineer
MRS. WALKER	60's, retiring school teacher
DOCTOR	Any adult age, Mary's physician
DUTCH KINDELBERGER	60's, owner of North American Aviation
PAUL MORSKY	30's, an employee recruiter
FATHER MACKEY	50's, a Catholic priest
DAVID SHELBY	Any adult age, a magazine reporter
PARK RANGER	Any adult age, a female Yellowstone ranger

The roles of Betty Manning and Mrs. Walker could be played by the same actor.

The roles of General Medaris, Michael Sherman, Dutch Kindelberger, and Father Mackey could be performed by the same actor.

The roles of Colonel Wilkins, the Doctor, Paul Morksy and the Yellowstone ranger could be performed by the same actor.

Other casting options: the roles of the doctor and ranger can be either sex, and so they could be performed by the same actor that plays Betty Manning and/or Mrs. Walker.

---

[1] Pronounced kuh-NAIR-ek
[2] Pronounced meh-DAIR-uhs

List of Scenes

Author's Note

How many people can say they were the son or daughter of not one, but two, rocket scientists? Growing up, my siblings and I were clueless as to how atypical our family was; we thought we were normal. Doesn't everybody discuss physics and chemistry at the supper table?

Though our parents were happy to share their scientific knowledge with us, the one thing they avoided as much as possible was sharing their personal family histories and backgrounds. Like their top secret aerospace work, their family histories, as it turned out, had a few events and incidents that caused them to become "top secret" as well.

This play tells a few chapters in the life of my mother, Mary Sherman Morgan, America's first female rocket scientist, and how she changed the course of aerospace history. I began writing it soon after her passing in 2004. How does one write a play about a parent whose entire life has been buried in riddles and secrets? It's not easy, and it took a great deal of research, including many interviews with my father, and the few co-workers who are still with us.

To them, I am eternally grateful.

George D. Morgan

THE SET: Most of the play takes place at North American Aviation's Office of Research and Development in Southern California. A half dozen steel desks, steel file cabinets, old chairs on rollers. Everything in those scenes should adhere to 1950's America. To one side is an open but separate office for the Research Department's manager - Tom Meyers. This office is also used for the General Medaris scene.

PROLOGUE

LIGHTS UP on the obit office of the LA Times. Barbara - head of the department - sits at a desk. Carley Morgan enters.

CARLEY: Excuse me - I'm looking for the obituary department.

BARBARA: May I help you?

CARLEY: Are you - Barbara?

BARBARA: Yes - what do you need?

CARLEY: My name is Carley Morgan - I sent you an obituary on my grandmother - Mary Sherman Morgan.

BARBARA: Oh yes - I remember.

CARLEY: I've been watching the paper, and I notice it still hasn't run.

BARBARA: I'm afraid we won't be able to print your your grandmother's obituary.

CARLEY: Why not?

BARBARA: Well, frankly Miss Morgan you make some wild claims in your piece - that she was America's first female rocket scientist, that she invented the rocket fuel that powered America's first satellite into orbit...

CARLEY: That's all true. What's the problem?

BARBARA: The problem is we can't independently verify
         any of it. There's nothing in the aerospace
         historical record that shows your grandmother
         ever existed - let alone that she accomplished
         anything.

CARLEY: Barbara - if my grandmother never existed, I
         wouldn't be here talking to you.

BARBARA: I'm sorry. Rules are rules.

CARLEY: This is absurd! You printed a half page
         obituary on that guy who invented the Oscar
         Mayer weenie whistle!

BARBARA: Of course - we were able to independently
         verify it.

CARLEY: What about all the people who accomplish
         great things every day, but who aren't in
         some historical record - are they just
         supposed to vanish into obscurity?

BARBARA: I think you're beginning to understand.

CARLEY: Understand? I'll tell you what I understand:
         In the 1950's my grandmother single-handedly
         saved America's space program, and nobody
         knows about it but a handful of old men.
         Here's what I'm gonna do: I'm going to find
         those men - I'm going to track them down, and
         I'm going to interview them. I'll get sworn
         affidavits, and you'll have your verification.

She turns and exits.

LIGHTS X-FADE to a card table.

Five men and one woman enter from various directions.
There is an almost ethereal element to their entrance.
They walk to stage front and face the audience.

BILL: My name is Bill Webber. I am an engineer.

JOE: My name is Joe Friedman. I am an engineer.

IRVING: My name is Irving Kanarek. I am an engineer.

DON: My name is Don Jenkins. I am an engineer.

TOM: My name is Tom Myers. I am an engineer.

MARY: My name is Mary Sherman. I am an analyst.

BILL: Every morning and afternoon...

JOE: ...we design the biggest and best rocket
          engines in the world.

IRVING: But at lunch time...

MARY: ...we play bridge.

Mary, Joe, Bill, and Don sit at a card table and pick
up their already-dealt cards. Irving and Tom stand
over to kibitz.

IRVING: Play the club.

LIGHTS FADE TO BLACK

END PROLOGUE

SCENE 1.

LIGHTS UP on the office of U.S. Army General John Bruce
Medaris. The general - in full uniform and wearing
reading glasses - is at his desk.

There is a bucket of ice on his desk with several
bottle of Coke. An open bottle is in his hand. It's a
hot day - he takes a sip.

There is a KNOCK from an O.S. door.

GENERAL MEDARIS: Come.

Miss Biddle - his secretary - enters.

MISS BIDDLE: Colonel Wilkins is here, sir.

GENERAL MEDARIS: Send him in. And Miss Biddle — I'd
like you to stick around for this.

She stands in place. As Colonel Wilkins enters, the
general removes his reading glasses.

COLONEL WILKINS: You needed to see me, sir?

The general pulls a Coke from the bucket.

GENERAL MEDARIS: Coke?

COLONEL WILKINS: No thank you, sir.

The general returns the soda to the bucket.

GENERAL MEDARIS: Colonel Wilkins; what do you know
                   about women?

COLONEL WILKINS: I know them when I see them, sir.

GENERAL MEDARIS: Good answer.
                   Here's one thing I know: there's no
                   such thing as being sort-of pregnant.
                   You're either pregnant, or you're not
                   pregnant — there is no in between.

COLONEL WILKINS: Yes sir - that's always been my
                   understanding.

GENERAL MEDARIS: Now - how much do you know about
                   rockets?

COLONEL WILKINS: I know they usually blow up.

MISS BIDDLE: Not Army rockets.

GENERAL MEDARIS: Not Army rockets.

COLONEL WILKINS: No sir — not Army rockets.

GENERAL MEDARIS: Colonel - have a Coke.

General Medaris hands Colonel Wilkins a Coke. This
time he takes it.

GENERAL MEDARIS: Now Colonel — what do you know about
Coke?

He is unsure what to say. Miss Biddle helps him out.

MISS BIDDLE: It's the pause that refreshes.

COLONEL WILKINS: It's the pause that refreshes.

GENERAL MEDARIS: Good answer. Now, rockets are like a
couple of Coke bottles.

COLONEL WILKINS: In what manner, sir?

General Medaris grabs another Coke bottle and holds it
above the one he has been drinking.

GENERAL MEDARIS: Two propellant tanks — one stacked on
top of the other. In one is the
oxidizer - in the other is the fuel.

MISS BIDDLE: You're losing him, sir.

The general sets the bottles down.

GENERAL MEDARIS: Do you own a car, Colonel?

COLONEL WILKINS: Yes sir — a fifty-seven Pontiac.

GENERAL MEDARIS: Did you buy it with the big engine?

COLONEL WILKINS: Of course — I'm an American; I always
purchase as much power as I can
afford.

GENERAL MEDARIS: Good. Now - think of your car: it
runs on two chemicals — a fuel and an
oxidizer. The fuel...?

COLONEL WILKINS: Gasoline.

MISS BIDDLE: He's brilliant, sir.

GENERAL MEDARIS: And the oxidizer...?

Colonel Wilkins has to think about this one.

COLONEL WILKINS: The...air. The air that goes into
                   the carburetor.

MISS BIDDLE: It's the oxygen in the air that
                   reacts with the gasoline.

COLONEL WILKINS: Sir - does she have to be here?

GENERAL MEDARIS: Shut up and listen - you might learn
                   something!
                   The oxidizer allows the fuel to burn,
                   and in the right proportion it helps
                   the fuel burn hotter, so you get more
                   energy. Fuels and oxidizers,
                   Colonel. Fuels and oxidizers.
                   If the American people knew how much
                   of their lives were governed by fuels
                   and oxidizers, they might crack open
                   a book and study once in a while.
                   Ya know what I mean?

COLONEL WILKINS: Yes sir.

GENERAL MEDARIS: Now - a rocket is exactly the same as
                   your car. It burns a fuel...
                        (points to one bottle)
                   ...and an oxidizer.
                        (points to second bottle)
                   Understand?

COLONEL WILKINS: Yes sir.

The general takes another sip of his Coke.

GENERAL MEDARIS: Have you read the intelligence report
                   we just received from Washington?

COLONEL WILKINS: The one about the Russian space
                   program? Yes sir.

GENERAL MEDARIS: Then you know the Russians are one
                   tit suck away from putting the
                   world's first satellite into orbit.
                   Do you know what that means, Colonel?

COLONEL WILKINS: Not exactly, sir.

GENERAL MEDARIS: It means America's space program is
                   years behind the Russians. It's
                   embarrassing! It's worse than
                   embarrassing; it's...it's...

MISS BIDDLE: Humiliating.

GENERAL MEDARIS: It's humiliating!
                   Our boys out at Huntsville - Doctor
                   von Braun's group - could have had a
                   satellite in orbit months ago. Do
                   you know why they haven't?

COLONEL WILKINS: I assume it has something to do with
                   politics.

GENERAL MEDARIS: Of course it has something to do with
                   politics! Everything involving
                   failure has politics at its
                   foundation!

MISS BIDDLE: The President keeps funneling our
                   budget to the Navy program.

GENERAL MEDARIS: Exactly. Damn swabbies. The
                   Vanguard - and those keep blowing up.
                   Fortunately the mucky-mucks in
                   Washington have finally gotten
                   their heads out of their asses and
                   realize they need the Army to come to
                   their rescue.

MISS BIDDLE: Go Army!

GENERAL MEDARIS: Now the government's asking us -
                   hell, they're begging us - to get von
                   Braun's Redstone rocket prepped and
                   ready for an orbital mission A-SAP.

COLONEL WILKINS: Amazing how public pressure helps
                   elected officials think more clearly.

GENERAL MEDARIS: I got a call from Doctor von Braun
                   himself this morning. He says his
                   engineers have calculated that the
                   Redstone, as currently configured,
                   has what it takes to get 95% of the

way into orbit. Do you know what 95% gets you in the satellite business?

COLONEL WILKINS: No sir.

MISS BIDDLE: It gets you bupkis.

GENERAL MEDARIS: It gets you bupkis!
                    (demonstrates with Coke bottle)
                    The rocket will keel over at a very high altitude and nose dive into the ocean – another embarrassing, humiliating failure for the U.S. of A.

COLONEL WILKINS: Yes sir.

GENERAL MEDARIS: Sometime in the near future Americans are going to open their morning newspaper and read a story about the Russians putting a satellite into orbit. You know what Mr. and Mrs. John Q. America are going to find out on that fateful morning, Colonel?

COLONEL WILKINS: No Sir.

GENERAL MEDARIS: They're going to find out that America's space program is "sort-of pregnant!" Do you believe it's possible to be sort-of-pregnant, Colonel?

Miss Biddle hands a folder to the Colonel.

MISS BIDDLE: Trust me – the answer is "no."

GENERAL MEDARIS: I'm sending you out to California - a little backwater they call Canoga Park. There's a company - North American Aviation - got some of the best engineers in the business.

MISS BIDDLE: Not as good as the Army.

GENERAL MEDARIS: Not as good as the Army. We're giving them the contract. Their charge is this: To come up with a new propellant combination that we can use in the Redstone to get that extra five per cent of performance, without changing one nut or bolt in the rocket's design. And we need it two days before yesterday.

COLONEL WILKINS: Yes sir.

GENERAL MEDARIS: Dismissed.

An exchange of salutes. Miss Biddle and the colonel move to exit.

GENERAL MEDARIS: Colonel.

The colonel turns around as Miss Biddle exits.

GENERAL MEDARIS: Tell those white-shirts at North American I want them to put their best man in charge of this project. Whoever they choose doesn't know it yet, but he's about to become the most important person in America.

COLONEL WILKINS: Yes sir.

Colonel Wilkins exits. General Medaris puts his reading glasses back on, takes a sip from his Coke, and goes back to work.

LIGHTS FADE TO BLACK.

END SCENE 1.

SCENE 2.

In complete darkness a male VOICE — devoid of emotion — is heard over a loudspeaker.

VOICE: LOX tank vent valve closed. Alcohol vent valve closed.
          (two beats)
       LOX tank pressure up. Alcohol tank pressure up.

                    (two beats)
          Turbo pump lube pump on. Turbo pump
          vent chamber open.
                    (more beats)
          Main LOX valve closed. Main fuel valve
          closed. Ninety seconds to ignition.

LIGHTS SLOWLY UP inside the control blockhouse of
rocket test stand #2 — Santa Susana Field Laboratory —
the hills above Simi Valley, California.

A cheap desk - and not much else. Two men in their
twenties - Bill Webber and Irving Kanarek - face the
audience with binoculars glued to their eyes.

They wear white dress shirts, dark ties and dark
slacks. I.D. badges are clipped to their breast
pockets. Crew cuts are the only choice on the menu.
These are the nerds that roamed the Earth before the
word was invented. Their clothes are slightly rumpled:
behold the world prior to "Wash and Wear."

VOICE: LOX and fuel tank regulators set to
          fifteen PSI.
                    (couple beats)
          Fire Ex system pressurized. Fire Ex main
          valve closed. Fire Ex pump on.

IRVING: Mary — you ought to come and see this; they're
          about to fire up the Navaho.

No response.

BILL: Are you coming or not?

VOICE: Peroxide regulator set to 450 PSI.
          Sixty seconds to ignition.

IRVING: Mary — you really ought to see this.

BILL: Hey — there's somebody laying on the ground
          next to the test stand.

IRVING: Looks like George Toumey.

                         HS 16

Mary Sherman — wearing a home sewn skirt and blouse circa 1950 — enters running. She grabs a third pair of binoculars off the desk.

BILL: That boy's in for a helluva wake-up call.

On a rotary phone, Mary dials a single digit number.

VOICE: Open fuel tank pressurizing valve. Open LOX
          pressurizing valve. Forty seconds.

As Mary talks on the phone, Don Jenkins enters. He grabs binoculars from one of the other two men.

MARY: This is Mary Sherman in blockhouse two.
          We need some technicians out at the test
          stand. Tell 'em their buddy George Toumey's
          been sampling the alcohol again.

BILL: Drunk as a skunk.

MARY: (still on phone) I warned him if he did this
          again we wouldn't stop the test. Now get him
          outta there!!

She hangs up the phone and steps to stage front — putting binoculars to eyes.

IRVING: Look at those techies run.

Joe Friedman enters and joins the crowd. He has a pair of binoculars and raises them to watch.

MARY: Pack of Winstons says they get knocked on
          their butts.

BILL: All I got are Kents.

VOICE: Igniter on. Twenty seconds.

IRVING: This is the moment the techies wish George had
          stayed on his diet.

BILL: Drag him — don't carry him!

IRVING: That's the way.

BILL: Oops!

IRVING: Lost a shoe.

MARY: They're still too close.

BILL: Here we go...

IRVING: A four thousand degree ass kick!

VOICE: Ten seconds...

BILL: Superman should be swooping down any second!

VOICE: Five...four...three...two...ignition start.

BRIGHT FLICKERING ORANGE LIGHTS flood the players as
they watch through binoculars. This is a static test
of an engine — not a rocket launch — so the binoculars
never move.

A beat — then the SOUND (which travels slower) reaches
the blockhouse with a continuous ear-splitting roar.

Everything in the blockhouse vibrates. Objects fall
off of whatever they're sitting on. It's like the
epicenter of an 8.0 earthquake.

The roar goes on and on. It is so loud there is a
warning printed in the play program for the benefit of
older audience members who must re-adjust their hearing
aids.

After about thirty seconds —

ORANGE LIGHTS CUT OFF. ROCKET SOUND EFFECT CUTS OFF.

VOICE: Engine cutoff.

The male engineers lower their binoculars, then start
to laugh, and their laughter builds as they exit.

BILL: That was great!

Now alone — Mary slowly lowers her binoculars. For a
moment she is in her own world.

MARY: Drunks.

As LIGHTS FADE TO BLACK - a LOUD KNOCKING begins. A
few beats - then the KNOCKING is heard again. Mary
turns in the direction of the sound.

END SCENE 2.

KNOCKING continues.

SCENE 3.

LIGHTS UP in the North Dakota farmhouse of Michael
Sherman - eighteen years earlier.

Mr. Sherman sits collapsed in a chair. Empty wine
bottles lie here and there.

The loud KNOCKING wakes him up and Michael stirs. The
KNOCKING continues - more insistent this time.

MICHAEL: Uhh - go away.

BETTY: Mister Sherman - it's Betty Manning from
          Social Services.

MICHAEL: I said go away.

The creak of a never-oiled door is heard.
Betty Manning enters.

BETTY: Mister Sherman - I've been knocking on your
          door for five minutes.

MICHAEL: Why didn't you just assume I wasn't home?

BETTY: I could see you through your front window.

MICHAEL: You're clever.

BETTY: Mrs. Walker tells me your daughter still
          is not attending school. Mary is nine years
          old - she is three years behind. Contrary to
          our agreement you still have not registered
          her for class.

MICHAEL: I need her here.   Somebody has to clean the
        creamer.

He searches for a wine bottle that isn't yet empty.

BETTY: I told you last week that you were being
         given your final warning.

MICHAEL: This is a farm - we have chores.

BETTY: This is a country - we have laws.

Michael's disdain of the government's interference with
his life sobers him up slightly.

MICHAEL: You government people wanna control
         everybody's lives! You made all my kids go to
         school - my sons, my daughter Elaine -
         everybody. They're all gone. There's no one
         left to work the farm but Mary. Do you
         paper pushers have any idea how much work
         it takes to operate a farm!? I never went to
         school - I turned out just fine. You're
         always telling me how to run my life, run my
         family. Let me tell you somethin' - this is
         my family - my house - my farm - my property!
         I make the decisions here!

BETTY: The sheriff is waiting outside, Mister
         Sherman.

MICHAEL: The river - what about the river? How's she
         gonna get across?

BETTY: We've already been over this. The State of
         North Dakota will provide a horse she can ride
         each morning across the river to school.

MICHAEL: (despondent) I need her. I just can't seem to
         get any work done by myself anymore.

BETTY: I brought the registration forms with me.

She hands him pen and paper.

BETTY: If you do not enroll her today you will be
         arrested.

HS 20

He signs the papers, then shoves them back.

BETTY: I'll have the horse sent over this afternoon.
          We will expect to see Mary in class tomorrow
          morning, eight o'clock sharp.

She moves to exit.

MICHAEL: Now who's gonna clean the creamer!?

BETTY: Sir - why don't _you_ give it a try.

Betty exits.

LIGHTS FADE TO BLACK

END SCENE 3.

SCENE 4.

LIGHTS UP on the Office of Research and Development of
North American Aviation. A half dozen desks and
chairs. Every desk has a desk lamp, papers, file
folders. It is lunch time. The engineers are in the
middle of a bridge game.

DON: I hear we're getting a visitor from White
          Sands.

BILL: Some new contract supposedly.

MARY: I hope it's a new engine - we could really use
          the extra work.

JOE: Could be. The Navaho Project's winding down.

IRVING: White Sands. That means they'll be sending
          a government goon.

BILL: Those "government goons" pay your salary,
          Kanarek.

Don notices off stage someone approaching.

DON: Uh – ex-nay on the government-say.

Colonel Wilkins enters, wearing Ray-Bans and carrying a
thick binder.

COLONEL: I'm looking for the Engineering Department of
          the Office of Research and Development.

BILL: You found it.

COLONEL: I found it. This is the engineering
          department.

JOE: This is it.

COLONEL: Could I ask what on earth you're doing?

IRVING: Playing bridge.

MARY: So what kind of a contract are you bringing
          us?

COLONEL: My mission here is top secret.

DON: Hey – I got a question about that. Could
          you explain to us: What exactly is the
          difference between secret and top secret?

JOE: Yeah – how much more secret than secret is
          top secret?

IRVING: If something is top secret, can it really
          be any more secret than just regular secret?

BILL: How exactly do you measure degrees of
          secrecy?

IRVING: If secrecy has different degrees, there must
          be some way of measuring those differences.

DON: Exactly. Exactly. And if there's
          measurement, there must be units – like
          meters, nano-seconds, gighertz.

JOE: Gigahertz of secrecy.

IRVING: Now we're getting somewhere.

HS 22

COLONEL: I have an appointment with Mister Tom Meyers.

BILL: Right through that door.

The colonel heads toward Tom's office door.
Mary shakes her head, smiling.

MARY: Gigahertz of secrecy. You boys are bad.

The engineers chuckle.

The colonel returns. He looks at one of the
engineers and points to Mary.

COLONEL: Have that secretary bring me a cup of
         coffee.

He moves towards Tom's office.

MARY: (gathering the cards) My deal.

The colonel enters the office of Tom Meyers – manager
of Research and Development at North American
Aviation. Charts, graphs and aerospace photos line
the walls. A golf bag with clubs sits in a corner.

COLONEL: I'm looking for Tom Meyers.

TOM: Colonel Wilkins?

They shake hands.

COLONEL: I met a few of your engineers coming in
         here. They're all playing bridge.

TOM: It's lunch time.

COLONEL: (checks his watch) So it is.

TOM: You mentioned on the phone the Army wanted to
         discuss a new contract.

COLONEL: Yes. As you know, the Redstone rocket program
         has been a terrific success.

TOM: Of course. We built the booster.

COLONEL: Which is why I'm here. The Army's been given
        the green light to adapt the Redstone for an
        orbital flight.

TOM: Well it's about time! Frankly everyone around
        here is tired of being on a leash. What can
        we do to help?

The colonel hands a file to Tom.

COLONEL: We need to increase the Redstone's performance
        by at least five per cent - without changing
        any of the hardware.

TOM: In other words, the only things we can change
        are the propellants.

COLONEL: Precisely. And we need it three days before
        yesterday.

TOM: Von Braun's people couldn't solve this?

COLONEL: Unfortunately, no. (opens the binder) I've
        been going over North American's list of
        engineers and I notice Irving Kanarek works
        here. As the inventor of inhibited red fuming
        nitric acid - a propellant we've had great
        success with on the Nike - Kanarek has a blue
        ribbon reputation with the Army. He seems to
        be the most qualified candidate. I'd like you
        to put him in charge of this.

TOM: Well, Irving's a good engineer, but
        theoretical performance calculations are not
        his specialty.

COLONEL: You've got nine hundred engineers on your
        payroll. I've read the background bio on
        every one of them. I believe Kanarek's our
        boy.

TOM: I disagree.

COLONEL: Well who would you choose?

TOM: When you're talking about theoretical
           performance calculations there's only one
           choice around this place: Mary Sherman.

COLONEL: Mary. As in M-A-R-Y Mary?

TOM: That's right.

COLONEL: You have a female engineer?

TOM: Actually, she's an analyst.

COLONEL: An analyst.

The colonel pages through the bios.

COLONEL: I don't have her in my book.

TOM: She's probably in the miscellaneous section.

The colonel finds Mary's bio in the back of the binder.

COLONEL: Ah - Mary Sherman.

He quickly reads her short bio.

COLONEL: According to this, Miss Sherman attended
           DeSalles College in Ohio for one year, then
           dropped out to work in a weapons plant before
           being hired here. No doctorate - no masters;
           not even a bachelors degree. No noteworthy
           successes or accomplishments - no published
           papers. Looks like the only thing we can say
           about her is she has a high school diploma
           from Ray, North Dakota - wherever that is.

TOM: All of that's true, however...

COLONEL: I'm sure you know a lot more about rocket
           science and technology than I do, but General
           Medaris told me to make sure you put your best
           man on this project, and I don't think he was
           speaking figuratively!

The Colonel turns pages in the bios.

COLONEL: Now, Mister Kanarek on the other hand...

Tom opens his door and steps out onto the engineering floor.

TOM: Hey Bill — ya got a minute?

BILL: Sure.

Bill enters Tom's office.

TOM: I want you to meet Bill Weber — one of my best engineers.

Bill and the Colonel shake hands.

Mary waves Joe over to take Bill's place at the card table.

MARY: Joe.

As Joe sits at the card table —

TOM: Bill — the Colonel here needs the very best theoretical performance person we've got for a special project. Who would you recommend?

BILL: Theoretical performance? I'd recommend Mary.

COLONEL: But Irving Kanarek has a much better...

BILL: Kanarek!? Oh please. Irving couldn't calc his way out of a toaster.

Again Tom steps out and addresses Don.

TOM: Don! Can I see you?

Don leaves the card table and enters Tom's office.

MARY: I'll play both hands.

TOM: A specialist for a project involving theoretical performance calculations. Who would you choose?

DON: Hmm. I guess...probably Mary.

COLONEL: Are you familiar with Irving Kan...

TOM: (calling from doorway) Irving!

Irving stands up.

IRVING: Sounds like the military-industrial complex
          is calling.

Irving enters Tom's office.

Mary and Joe now each play two hands.

TOM: Colonel – meet Irving Kanarek. Irving – the
          Army needs the very best theoretical
          performance person we've got for a special
          project. Who would you recommend?

IRVING: I could do that.

BILL: Irving – get real...

IRVING: I could!

Bill rolls his eyes.

IRVING: Hey – I invented inhibited red fuming nitric
          acid – remember?

BILL: (super mockery) I invented inhibited red-
          fuming nitric acid. I'm such a pretty boy!

TOM: Bill...

BILL: You know perfectly well theoretical
          performance is Mary's department.

COLONEL: Perhaps North American isn't the right company
          for this job.

The engineers realize they are close to losing a
contract. They – sans Tom – go to the door and shout:

ENGINEERS: Mary!

MARY: Let's see what the boy's club is up to.

Mary leaves Joe alone at the card table and enters
Tom's office.

TOM: Von Braun needs a new propellant combination
           to boost performance on the Redstone. Any
           idea who I should give that to?

MARY: Well, you would need somebody with experience
           doing theoretical performance calculations.
           I'm sure Irving could handle that.

COLONEL: There. Someone's finally making sense.

TOM: Colonel - meet Mary Sherman. Mary - Irving
           thanks you for your humble brown-nosing. Now
           honestly - who would you really pick?

MARY: Oh - you want honesty. Well in that case if
           you give that project to anyone but me I'll
           kick your ass. You know perfectly well that
           theoretical performance is my specialty. I
           did all the propellant work on the NAVAHO -
           the NALAR program was all my work - and let's
           not forget...

TOM: It's okay - it's okay.

BILL: (to the colonel) She's pretty good
           at the honesty stuff.

TOM: Why don't you all go back to work.

The engineers exit Tom's office.

DON: (to Mary) You do know that guy's your boss.

Lunch time is over. They put away the card table and
go back to work.

The Colonel just won't let it go.

COLONEL: A high school diploma! A high school diploma,
           for godsakes! Nine hundred engineers - all
           male - all with college degrees - and you're
           telling me the person best qualified to do

this job is not just the only woman engineer in the company, but also the only engineer without a college degree!? That's what you're telling me?

TOM: She's not an engineer, she's...

COLONEL: ...an analyst - yes, you said. She has no résumé. No track record. She's never accomplished anything noteworthy!

TOM: Colonel; no one accomplishes anything noteworthy - until they do it for the first time.

LIGHTS FADE TO BLACK.

END SCENE 4.

SCENE 5.

SPOT ON an empty part of the stage. The voice of Mary's high school teacher, Mrs. Walker, is heard.

WALKER: (out of light) Mary? Where are you?
          (beat or two) Mary - your family and friends are here.

An elderly MRS. WALKER enters the spot.

WALKER: Mary? Are you out here?

MARY: (out of light) I'm here.

Mrs. Walker takes several steps - and the SPOT FOLLOWS her. She and the FOLLOW SPOT find Mary - sitting in a chair and wearing a graduation cap and gown.

WALKER: This is your graduation party. You should be inside.

MARY: I wanted to watch the stars.

Mrs. Walker looks up for a moment - but fails to see the priority.

WALKER: Well yes – they're nice.
         My, you've been thoughtful lately.

MARY: I've been thinking about my future.

WALKER: Mary – a girl as smart as you has her future
         already set. You'll be married, have
         children, and – now that I'm retiring – take
         over as Ray's new school teacher.

MARY: Married? Maybe. Children? Doubtful.
         I want to be around people of intellect.
         That leaves out children – and it certainly
         leaves out Ray.

WALKER: What are you saying?

MARY: There are plenty of others out there who
         would make great school teachers.

WALKER: I submitted your name to the Board. They
         already approved it.

MARY: You should have discussed that with me.

WALKER: Everyone in Ray has always assumed you would
         be our next teacher.

MARY: They assumed wrong. I'm leaving Ray, and
         I'm leaving North Dakota.

WALKER: We're not good enough for you?

MARY: Sure you are.

But we sense she is lying to be polite.

WALKER: Who's going to take my place?

MARY: It won't be me – I have other plans.

WALKER: Other plans? You've never mentioned anything
         about other plans.

MARY: I'm leaving tonight. This will be our last
         chance to say good-bye.

HS 30

They briefly hug.

WALKER: Do your parents know about this?

MARY: They will in the morning.

WALKER: Where will you be going?

MARY: I've enrolled in DeSalles College - in Ohio.

WALKER: I've never heard of it.

MARY: It's not the greatest college in the world,
           but it's all I can afford. Their science
           program looks pretty good; I've decided to
           become a chemist.

WALKER: A chemist! Mary - girls don't become chemists
           or scientists. That's a man's career.

MARY: Don't you think I can do it?

WALKER: Just because we can do something doesn't mean
           we should.

MARY: Well you see, that's where we differ.

As she walks away -

LIGHTS FADE TO BLACK

END SCENE 5.

SCENE 6.

LIGHTS UP on the Office of Research and
Development.

Irving is at his desk - alone. Joe Friedman enters
carrying a rolled set of blueprints and goes to his
desk. He opens the blueprints.

A beat or two – then Don Jenkins enters with a small vase of flowers and places them on Mary's desk.

DON: Hey Joe – how's that camera thing
         coming along?

JOE: It's called a "lens," Jenkins.

IRVING: Better not let Tom see you're working
         on it on company time.

JOE: I'm not doing it on company time.
              (a beat or two)
         Maybe once in a while.

DON: Joe, here, is gonna be a rich inventor some
         day. Isn't that right, Joe.

Bill enters with a coffee mug and goes to his desk.

BILL: Tom alert.

JOE: Jenkins – you wouldn't know a great invention
         if it flew up your ass.

Don has gotten Joe's goat. He and Irving chuckle over it.

BILL: By the way – I hear Tom's hired a new heat
         transfer specialist.

DON: Heat transfer? Am I being replaced?
         I'm being replaced, aren't I?

Tom Myers enters with the new recruit – Richard Morgan. He is holding a brown cardboard box filled with his personal belongings. He is dressed like the other men – same crew cut hair – but this hair is bright red.

TOM: Everybody – as you know we've been looking to
         hire someone to handle the heat transfer
         problems on the NAVAHO. Fresh out of Caltech –
         I'd like to introduce Richard Morgan. Joe,
         he'll be working with your group.

JOE: We can use you.

HS 32

DON: I can do heat transfer.

TOM: That's great, Don.

Tom puts a cheap-looking name plate on a desk.

TOM: This will be your desk. Any questions?

RICHARD: Just one. I had to get a top-secret clearance
          to work here. What do I tell people when they
          ask me what I do?

TOM: Boys and girls - what do we say?

ALL: A little of this - a little of that.

TOM: Good luck.

Tom leaves. Richard gets comfortable at his new desk.

Mary enters from her break holding a mug of coffee. She
notices the vase of flowers and moves them to a corner
of the desk.

MARY: Thank you, Don. (notices Richard) Do we have a
          New Guy? (checks his name plate)
          Richard Morgan.

DON: So what's your story, New Guy?

BILL: Yeah - who are you and why should we care?

MARY: What they mean is: what noteworthy
          accomplishments do you have that would make us
          kneel down and worship at your feet?

RICHARD: Well - I played college football in the Rose
          Bowl.

DON, JOE, IRVING, BILL: WHOA!!

The men jump up from their desks, kneel down, and
bow to Richard, chanting:

DON, JOE, IRVING, BILL: We're not worthy...we're not
worthy...

MARY: Get up – you're embarrassing! (they comply)
           You. You played college football in the
           Rose Bowl.

RICHARD: More than once.

MARY: More than once. Pack 'o Winstons says New
           Guy has never even seen the inside of the
           Rose Bowl.

BILL: All I got are Kents.

MARY: Prove it.

RICHARD: Well – you could call my old coach; Burt
           Labrucherie.

MARY: Don't tell me – he's out of the country on
           vacation – or he's dying.

RICHARD: He's retired. You could probably reach him
           at home.

Mary picks up the rotary phone receiver on her desk and
dials a single number: 9.

MARY: I need an outside line. (to Richard) What's
           the number?

RICHARD: Dickens five eight seven four two.

MARY: Dickens five eight seven four two.
           (a beat or two) Hello – May I speak to Burt
           Labrucherie? ...Burt? – this is Katherine
           Hepburn. I'm doing an employment background
           check on a Richard Morgan – do you know him?
                   (a beat)
           He put on his resume that he's played college
           football in the Rose Bowl...
                   (a beat or two)
           Is that right...Is that so...really.. that's
           interesting, I never knew that. Okay, sir –
           thank you very much.

She hangs up.

HS 34

MARY: Ooooookaaaay. How many of you chuckleheads
           knew that New Guy went to Caltech?

The men, except Richard, raise their hands.

BILL: Tom told us.

MARY: And how many of you knew that Caltech plays
           all of its home games in the Rose Bowl?

The men, except Richard, raise their hands.

JOE: It's common knowledge.

Mary nods - realizing she has been snookered.

MARY: So, New Guy thinks he's clever.

RICHARD: New Guy is clever.

MARY: Richard. I guess that means I'll be calling
           you...Dick.

The men have a chuckle over this.

RICHARD: So what about you. Any noteworthy
         accomplishments?

MARY: Quite a few as a matter of fact.

RICHARD: Like?

MARY: I was...my high school's valedictorian.
           Class of '39.

RICHARD: A girl valedictorian? I don't believe it.

MARY: Gonna cost you to find out.

RICHARD: I don't smoke.

DON: We accept cash in lieu.

RICHARD: Okay. I got fifty cents says you weren't even
         in the top ten of your class.

MARY: Pick up the phone. We're going to call Ray
           High School, Ray, North Dakota.

He picks up his desk phone but is unsure what to do.

MARY: Dial nine.

A rotary phone – he dials a nine. He wonders what to
do next.

MARY: Wait for the impatient lady.

RICHARD: (a beat) Oh, Hi. I, uh...yes...uh...

MARY: You need an outside line.

RICHARD: I need an outside line...Thank you.

MARY: Area code seven zero one. Bismark three four
           six seven two.

RICHARD: Area code seven zero one. Bismark three four
           six...

MARY: Seven two.

RICHARD: Seven two.
               (a beat or two)
           Yes, this is Humphrey Bogart – I'm doing an
           employment background check on one of your
           former students – Mary Sherman...Class of '39,
           that's right. Miss Sherman claims on her
           application that she was your school's
           valedictorian.
               (a beat or two)
           Uh-huh...Is that so?...I see...That's very
           interesting...Thank you. And, uh...(Bogey)
           Here's looking at you, kid.

He hangs up.

BILL: You owe us each four bits.

RICHARD: Hey – I haven't told you the answer yet!

IRVING: We already know the answer.

HS 36

RICHARD: (to Mary) Ray High School only has seven
students. In 1939 you were the only
graduating senior, so you were...

BILL, DON, IRVING, JOE: ...the class Valedictorian.

DON: Don't feel bad - we all fell for that when we
first started working here.

Her dignity restored, Mary goes back to work.

With Tom gone, Joe has gone back to his blueprints.
Something about the blueprints catches Richard's eye.
He steps closer.

RICHARD: Is that a tri-focal lens?

Joe reacts like he has received an electric shock -
quickly rolling up his blueprints.

JOE: NO! No it's not!

RICHARD: Yes it is - I recognize the...

JOE: It's a two-speed reciprocal inverse reaction
formulator! It has nothing to do with
cameras or zoom lenses. What's your interest
in zoom lenses, anyway!? Why do you keep
talking about zoom lenses!?

RICHARD: A zoom lens.

JOE: Who are you!? What do you want!? Who are you
working for!?

RICHARD: North American Aviation - I thought.

JOE: A new guy suddenly appears out of nowhere and
he just happens to be curious about my design?

RICHARD: Look, I'm sorry.

DON: Aren't we supposed to be designing rocket
engines?

IRVING: Maybe it's a new product line.

Don and Irving have their chuckle.

Richard returns to his desk, but Joe keeps a wary eye
on him. Richard starts removing items from his
cardboard box. One of them is a new Nikon camera. Joe
notices it immediately — and we see the lust.

JOE: Is - is that the new Nikon SP Rangefinder?

RICHARD: Sure is.

JOE: How did you get it!? They're not supposed to
          be available for another three months.

RICHARD: After graduation I spent a couple of weeks in
          Japan. Picked this baby up at a store in
          Okinawa.

Oh the camera-lust.

JOE: Does it have the new titanium shutter
          curtains?

RICHARD: Yyyep.

JOE: With motordrive?

RICHARD: Of course.

JOE: A brightline illuminator? Single non-rotating
          speed dial? Six built-in framelines?

RICHARD: Color coded.

Joe regrets the way he treated Richard moments ago.
He's embarrassed, but camera-lust overpowers him.

JOE: Look — maybe I was a little rude a moment ago.

RICHARD: I think you were.

Joe ponders his options — then makes an offer.

JOE: I'll show you my blueprints if you'll show me
          your SP.

RICHARD: I don't know...

JOE: Don't make me beg!

Richard and Joe switch prizes - examining them closely.

RICHARD: This lens isn't big enough for a motion picture
            camera.

JOE: Look at the size of that RF window.

RICHARD: Wait a minute - you're indicating here a
            standard attachment thread size. Are you...
            are you designing a zoom lens for consumer use?

JOE: I figured out a way to make them lighter,
            smaller, and cheaper. The Japanese have been
            working on it for years - I solved all the
            problems they couldn't.

RICHARD: This is huge. Have you made a prototype?

JOE: I'm a designer - not a metal bender.

RICHARD: I have a little machine shop in my garage.
            I could help you make this.

IRVING: Five o'clock. I'm getting outta here.

Don, Bill and Irving start packing up for the day.
Joe hands the camera back. Richard hands him the
blueprints.

JOE: We'll talk some more.

DON: See ya tomorrow.

Don and Bill grab their briefcases, and begin to take
their exits with Irving. Bill notices Irving has
forgotten his briefcase, and points to it.

BILL: Irving.

Irving returns to his desk and grabs his briefcase.

DON: Screw that head on, buddy. (to Mary) You
            coming?

MARY: I've still got work to do.

Joe takes his blueprints and briefcase, grabs his coat,
and exits with the other engineers, leaving Mary and
Richard alone.

A beat - then Don peeks momentarily around a corner.
Then he leaves.

Most of the bright overhead fluorescent lights -
hundreds of them in the Research Building - are shut
down by maintenance for the day. The lighting becomes
subdued.

Though she initially has no interest in New Guy, as the
following dialogue progresses, their rocket engines
start warming up.

MARY: (boooring!) So you were a football player.

RICHARD: Yeah.

MARY: I'll bet you had dozens of sorority girls all
           over you.

RICHARD: Actually, I only played one year. Busted my
         knee and had to quit. And there are no
         sororities at Caltech; it's men-only.

MARY: Ah yes. Men only. I'm familiar with that
           environment.

Richard picks up the deck of cards on her desk.

RICHARD: You've been gambling since I got here.
         You play poker, too?

MARY: (grabs the cards) Only sailors and
           Mississippi harlots play poker.

She stows the cards in a desk drawer.

MARY: We play bridge. Every day at lunchtime.

RICHARD: Bridge. Are you any good?

MARY: Am I any good!?

RICHARD: Yeah. Are you any good?

It's an affront - but she decides to answer.

MARY: Took second place in the county tournament
          last year.

RICHARD: If you had a better partner, maybe you would
          have taken first.

MARY: That's the excuse I've been using.
          Do you play?

RICHARD: Doesn't everybody?

With the knowledge that he is a bridge player, the
flower of interest in New Guy begins to blossom.

MARY: So what about you - are you any good?

RICHARD: I don't play at tournament level, but I do win
          a few hands now and then.

MARY: Perhaps with the right partner you could play
          some tournaments.

RICHARD: What do you do here - besides winning bets.

MARY: I'm an analyst.

RICHARD: An analyst. What's that?

MARY: An analyst is someone engineers go to when
          they're too embarrassed to ask their fellow
          engineers for help. An analyst is someone who
          performs work others get promoted for. An
          analyst is someone who does twice the work of
          an engineer and gets paid half as much. An
          analyst is someone who wishes she could be
          judged strictly by her talents and abilities.

RICHARD: So you like your job.

MARY: The truth is, I do like it.

The distance between them begins to diminish.

RICHARD: What's keeping you from being an engineer?

MARY: A piece of paper.

RICHARD: What kind of projects do they have you on?

MARY: Theoretical performance calculations, mostly.

RICHARD: And that is...

MARY: Well, if you have two rocket propellants, and
          you want to know how well they'll perform...

RICHARD: Perform.

MARY: ...together...

RICHARD: Together.

MARY: ...without going to the time and expense
          of actually testing them, you come to me. It
          requires a lot of number crunching; the kind
          of grunt work that...

RICHARD: Grunt work?

Nervous - Mary knocks over the vase of flowers Don gave
her. She picks it up and tries to appear nonchalant.

MARY: The kind of...grunt work...that most engineers
          don't like to do.

RICHARD: So you work with very powerful rocket
          propellant combinations.

MARY: Yes. Combinations. Of two.

RICHARD: Two.

MARY: Two. As in a pair.

RICHARD: As in a fuel...

MARY: ...and an oxidizer. The oxidizer is very
          important.

HS 42

RICHARD: Of course - it makes the fuel burn...

MARY: ...hotter.

Tom enters - holding his suit coat and briefcase.

TOM: Mary -

Richard and Mary move apart as casually as possible,
looking in different directions.

TOM: I convinced the colonel you're the best person
           for the Redstone project. They're sending the
           contract over tomorrow. We'll start work on
           it immediately. Congratulations.

MARY: Thanks, Tom.

RICHARD: Good-night, Mister Meyers.

TOM: Call me Tom.

Tom looks at each of them - sensing that something is
going on - then he exits.

Now the two of them are really alone.

MARY: I have work to do.

She moves toward her desk. He interrupts her.

RICHARD: Mary...

MARY: Yes?

RICHARD: Would you...

MARY: Would I...

RICHARD: ...like to go out to dinner some time?

MARY: You mean - like a date?

RICHARD: Yeah. Like a date.

MARY: You're asking me on a date?

RICHARD: Sure. You've been on a date before.

MARY: Of course. Lots of them. Hundreds. Thousands.

RICHARD: Maybe you're too experienced for me.

MARY: No I'm not. How many dates have <u>you</u> been on?

RICHARD: A few.

MARY: Hm.

RICHARD: So what is your answer?

MARY: Sure. Dinner sounds great.

RICHARD: Great.

MARY: Great.

RICHARD: Saturday?

MARY: Saturday.

Richard grabs his coat and starts to leave.

RICHARD: Good, night - valedictorian. (turns to leave)

MARY: Good night - tight end.

Shocked at the unintended double entendre, she slaps
her hand over her mouth.

LIGHTS CUT TO BLACK

<u>END SCENE 6.</u>

<u>SCENE 7.</u>

LIGHTS UP on Tom's office.

Tom is alone - working at his desk.

Bill enters.

BILL: You wanted to see me?

TOM: Yeah, Bill - come on in.

Tom picks up a large manila envelope on his desk.

TOM: I just read the paper you and Mary are
          submitting for publication. It's very good.
          Top quality work.

BILL: So we can send it out?

TOM: Absolutely.

Tom hands him the envelope.

BILL: That's great.

TOM: Uh, however, you will have to make one small
          correction.

BILL: You found a mistake?

TOM: Yes. You identified Mary as an engineer.
          You can't do that - she's only an analyst.

BILL: She does the work of an engineer.

TOM: A stable boy can put on a band-aid; that
          doesn't make him a doctor.

BILL: That's not a fair comparison. She's every bit
          as good as the rest of us - and better than
          some.

TOM: Doesn't matter - company rules; no college
          degree - no title.

BILL: Tom - you're the one that put her in charge of
          the Redstone project. You obviously recognize
          her talents.

TOM: That's different. That's a project that
          just happens to be uniquely suited to her
          specialty.

BILL: If I can't list her as an engineer, I'm not

going to publish.

Bill hands the envelope back to Tom.

TOM: Bill - don't get me wrong - we want you to
         publish. It'll be a feather in the cap of
         this department and the entire company. It's
         groundbreaking work. We're going get
         contracts out of this. It's going to mean a
         lot of new work and a lot of money for
         everybody. You just have to change one lousy
         word. Change 'engineer' to 'analyst' and
         we're done.

BILL: Forget it. I've changed my mind - I'm not
         going to publish.

Bill leaves the office - entering the engineering area.
Mary is the only person there - working at her desk.
Tom follows Bill and escorts him back to the office.

TOM: Bill - please. Look; your paper is brilliant.
         It's going to be a career-maker for you.

BILL: I didn't write it.

TOM: What are you talking about?

BILL: (blows up) I said I didn't write it! What;
         are you deaf!?

A few quiet beats - these men are colleagues who never
argue.

BILL: Mary wrote it. She just put my name on it
         because I helped her out with some of the
         grunt work. I did some heats of formation
         calculations, a few other odds and ends. Yeah
         - I changed her title to engineer; that's one
         thing I wrote.

The phone on Bill's desk RINGS. Mary answers it.

MARY: Research...Just a moment.

Mary enters Tom's office.

MARY: Bill — you have a call.

BILL: And I'm not going to change it.

MARY: Change what?

TOM: We'll talk about this later.

BILL: I think we should talk about it right now.
          Let's talk, Tom. Let's tell everybody...

TOM: I said later!

Tom storms out of his own office and exits.

Bill and Mary return to the engineering area.

MARY: What was that about?

BILL: Just an argument over the Navaho injector
          manifold.

MARY: I thought we solved that problem.

They sit at their desks. Bill hangs up the phone
without talking to whoever was on the other end.
They return to work. During the quiet moments that
follow, Bill notices a change has come over his co-
worker.

BILL: Are you smiling?

MARY: No.

BILL: Yes you were. You were smiling.

MARY: I was not smiling.

BILL: What is it?

MARY: Nothing.

BILL: Uh-huh.

They go back to work. A few beats — then Mary starts
humming.

BILL: Oh my god you're humming.

MARY: Was I?

BILL: Yes - you were humming.

MARY: There's nothing wrong with humming.

BILL: I've worked next to you for five years.
          You have never hummed.

MARY: I just feel like humming today.

BILL: What is it?

No response - so he wheels his chair over to her desk.

BILL: Have you met somebody?

She just smiles.

BILL: Do you have a date?

She can't hide it any longer, and starts laughing.
She nods.

BILL: Oh my god - good for you. When?

MARY: Saturday.

BILL: Saturday! This Saturday?

She smiles and nods. Bill is ecstatic for her.

BILL: Anybody I know?

The smile gets a little bigger.

BILL: You're not going out with Don...

He can tell by her reaction it's not Don.

BILL: Well who is it?

She crooks her head slightly toward Richard's desk.

BILL: New Guy? You're going out with New Guy!?

HS 48

She smiles and nods.

BILL: You little sex maniac.

They laugh and hug.

BILL: I am so proud of you!

As Bill returns to his desk he starts la-la-la-ing the
Wedding March.

MARY: Bill!

LIGHTS CUT TO BLACK.

END. SCENE 7

CHURCH ORGAN MUSIC - the same wedding march - plays as
an interlude. It fades, and we hear laughter and the
clink of glasses.

SCENE 8.

LIGHTS UP on the Research Office.

A small office party. The desks have glasses, buckets
of ice, bottles of champagne. There are some
decorations. The engineers are busy preparing for the
party. Everyone but Don seems to be happy.

DON: I can't believe she ended up with New Guy.

IRVING: Better get over it, buddy.

As Irving steps away to help with party decorations,
Don grabs an open champagne bottle. He steps downstage
and performs a monologue, ignored by everyone else.
During the monologue Don pours and drinks and pours and
drinks, and pours and drinks....

DON: I coulda gone to Caltech and played football
          and busted my knee and been the New Guy.
          I know how to be a New Guy. I can be the

HS 49

handsome New Guy who invades a company like
John Wayne and the first day he's there he
takes all the women...all one of them.
I can do the John Wayne thing. I'm articulate
- I'm smart - I can do heat transfer. We
didn't need a new heat transfer specialist.
I studied thermodynamics - I know Fouriers
Law. Conductance equals K over delta X.
I can even pronounce Fouriers Law backwards:
Wall-Sreerowf - Wool-Seeroofy - Wil-
Seelysoolyruf. Whatever.
I was doing thermal conductivity calculations
in the fourth grade. I know all about
proportionality coefficients, phase changes,
Dittus-Boelter correlations.
Chicks dig that sort of thing.
It's the red hair, isn't it? I coulda been
born with red hair. I coulda changed my
genetic makeup. How was I supposed to know
she likes guys with red hair? Who am I,
anyway. Just some dumb schmuck who brought
her flowers every week.
I can be an inventor. I can make a new lens.
I can be the bad boy. I can wear tattoos and
ride a motorcycle. I can go down the rabbit
hole. I can be the Cheshire cat! Meeooow!

IRVING: (finally noticing Don) Have you been drinking?

DON: No.

Joe sees the newlywed couple approaching.

JOE: Here they come!

The engineers cheer as Richard and Mary - just married
and wearing their wedding clothes - enter. Mary wears
a bridal veil, which trails down her back.

Everyone except Don applauds and cheers their entrance.
Don raises his glass for a toast.

DON: To Mary and the New Guy.

Don swallows another glass of champagne.

RICHARD: How long does someone have to work around

HS 50

here before people stop calling him the New Guy?

JOE: Crumski – over in Fabrication; they called him New Guy for twenty-three years.

RICHARD: What happened?

JOE: He retired.

IRVING: A toast. May your propellants be exotic, and your fuel tanks be ever full.

GROUP: Hazah!

They all take a drink.

MARY: I have one. To my two favorite inventors; may they rocket to the toppet!

GROUP: Hazah!

They all take another drink.

RICHARD: Speaking of that – I brought the prototype.

From his tuxedo pocket, he hands Joe a rudimentary zoom lens – little more than a black tube – a simple prototype.

JOE: You finished it?

Joe replaces the lens on his camera with the new zoom lens.

MARY: (incredulous) You had that in your pocket the whole time – during the ceremony?

Richard pulls another lens from the other pocket.

RICHARD: Made one for myself, too.

Mary just rolls her eyes.

JOE: (overwhelmed) It's beautiful.

IRVING: So Richard - tell us about the honeymoon.

RICHARD: We're going camping in Yellowstone.

BILL: Camping!?

IRVING: Gee - that doesn't sound like much of a
          honeymoon.

Mary folds her arms and takes a defensive posture.

MARY: It was my idea.

IRVING: Oh - great idea!

BILL: Absolutely!

DON: Fantastic.

A quiet moment of reflection, then -

JOE: First photo with the prototype.

Richard and Mary pose. Don horns in next to Mary.
Joe takes their picture.

MARY: I suppose we should go.

They kiss. There's nothing more to say - and everyone
knows it.

Mary turns around and throws the corsage. Don catches
it. General laughter and hilarity. In that moment of
perfect happiness Tom Meyers shuffles in - head bowed,
one hand in pocket - the other hand holding a single
sheet of paper.

The room quiets down. Mary loses her smile.

MARY: Tom - what's wrong?

A few beats while he gathers his thoughts.

TOM: When I was a boy my father and I used to build
          these little model rockets. At least once a
          month we'd go out to a field near our house
          and shoot them off. Then we'd go back to the
          garage and try to figure out how to make them

fly higher and farther. "Never be satisfied with second best," he said. "Always work towards improvement. Always be better than everyone else at what you do, and you will be a success." Years later – when I left for college to study engineering – I reminded him of those pep talks, and I promised to follow his advice.

BILL: Tom?

TOM: I've pretty much kept that promise – until today.

Mary steps next to Tom and gently places a hand on his shoulder.

TOM: We just got a teletype from Washington.
          A half hour ago the Soviet Union placed the world's first satellite into orbit.

Absolute stunned silence – it's a crushing blow. Mary takes the paper from Tom and examines it.

TOM: I'll be in my office.

He moves to go, then stops and turns to Richard, pulling a thick white envelope from his suit pocket.

TOM: This is from the management team – a little something to sweeten the honeymoon.

Tom stuffs the envelope in Richard's inside breast pocket.

TOM: The rest of you can go home early. Take the afternoon off.

Tom exits into his office.

A stoic silence. Mary places the teletype on her desk.

MARY: Bill?

BILL: Yes, Mary.

MARY: Have we tried the amines, yet?

BILL: No. No, no no! No way! You need to leave —
          right now. You and your new family are what's
          important. Get going, Mary! Get going on
          that honeymoon! We can handle things here
          for a week.

Mary reaches into a pocket of her dress, removes her
glasses, and puts them on.

DON: Uh oh.

MARY: Don — could you get me the heats of
          formation and densities on ethelinediamine
          and diethylamine. Bill — I'll need the
          molecular weights, boiling points, vapor
          pressures. Irving, I'll need specific impulse
          figures using LOX. Richard...

The newlyweds exchange a look. She puts her arms
around him and they kiss.

MARY: ...could you check out the comparative mixture
          ratios for me?

Richard removes his tux jacket and hangs it on a chair.
Everyone grabs a reference book, or a slide rule, or
some other tool of the trade — and gets quietly back to
work.

Still in her wedding gown, Mary moves to the
chalkboard. She begins to write the chemical formula
for ethelinediamine: $C_2H_4(NH_2)_2$ as —

LIGHTS FADE TO BLACK

END SCENE 8.

INTERMISSION

SCENE 9.

LIGHTS UP on two chairs. Paul Morsky — well dressed,
and holding a clipboard with numerous papers - enters
and looks around. He checks his watch. He seems
impatient.

Mary enters. She is carrying a large school textbook.
Her hair is tied up in a style we have not yet seen.

MARY: Are you Paul Morsky?

PAUL: Mary - thank you for meeting with me.

He shakes her hand.

MARY: I have to get to my math class...

PAUL: This will only take a moment. Mary — I'm what
          you might call a recruiter.

MARY: A recruiter. You mean like for the Army?

He smiles, then hands her a business card.

PAUL: No no. I work for a company; a company that's
          looking to hire people with certain talents.

MARY: (reads card) Plum Brook Ordnance. Ordnance —
          as in explosives. What kind of explosives do
          you make?

PAUL: Oh, a little of this, a little of that.

MARY: How did you get my name?

PAUL: You were referred to us.

MARY: Referred by whom?

PAUL: One of your teachers.

MARY: Why me?

PAUL: The war has hurt this country, Miss Sherman;
            more than many people are aware. We have a
            lot of skilled jobs that are not being filled.

MARY: Like...?

PAUL: Like chemical engineers. There's a big
            shortage of trained chemists right now.

MARY: You're aware I'm just finishing my freshman
            year here.

PAUL: Oh yes - we know all about you. Your teachers
            give you glowing reports. You're at the top
            of your class in both math and science.

MARY: DeSales College is not exactly Harvard.

PAUL: You're modest. I like that. How would you
            like to skip college and go straight into your
            career? We'll hire you as a chemist - right
            now. Not quite the same pay scale as if you
            had a degree, but close.

MARY: You must be desperate.

Paul says nothing - but we get the message.

She gives him his business card back.

MARY: Thank you for the offer, but I want to finish
            school, first.

PAUL: I'm not done telling you everything.

MARY: Well I'm done listening.

She turns to go.

PAUL: Mary - your country needs you!

MARY: (turns back) Everybody needs me! It's been my
            lifelong problem. Do you have any idea how
            many cows I've milked, how many creamers I've
            cleaned, how many bales of hay I've tossed!?
            I've spent a lifetime doing things for other
            people. College is something I finally get to

do for myself! I've spent every day since I
was nine years old dreaming of this moment,
and you want me to throw it away!?

PAUL: (mood change) You selfish little girl. Do you
          know how many good, young American boys die on
          the battlefield each day!? If we don't bring
          this war to a close soon, Miss Sherman, our
          country could be looking at hundreds of
          thousands of casualties. Perhaps millions.
          Our boys are giving their blood - can't we
          give a little bit of our time?

He hands her the clipboard. She reluctantly takes it.

MARY: How long would this job last?

PAUL: One year, tops.
          Two years, tops.

MARY: I suppose I could continue my education taking
          night classes.

PAUL: Sure. You could do that.

But in his voice there is a hint of doubt.

MARY: One year - two at the most. You're sure.

PAUL: Absolutely. Scout's honor.

He hands her a pen.

MARY: What would I have to do?

PAUL: You'll need to apply for a top secret
          clearance - just a formality in your case.
          Aren't too many Nazi spies coming out of Ray,
          North Dakota these days. Takes about a month.
          After that, you show up for work - seven
          thirty in the a.m.

She considers a moment - then signs the form. He takes
the clipboard back.

PAUL: I promise - what you'll be working on will be
          a helluva lot more important than milking

cows.

He shakes her hand, then moves to exit.

MARY: Hey - you still haven't told me what I'll be
          doing.

PAUL: A little of this - a little of that.

Paul Morsky exits.

END SCENE 9.

Mary puts down the textbook, hangs her purse on the
chair, unbuttons the top button of her blouse, and
removes the tie from her hair.

SCENE 10.

A DOCTOR enters - writing something on a clipboard.
Mary is stoic.

DOCTOR: The good news is it's not a death sentence.
          You'll probably live well into your eighties.

MARY: How did I get it?

DOCTOR: It's genetic. You inherited it from one of
          your parents.

MARY: They've never mentioned anything.

DOCTOR: They may not even know. The symptoms aren't
          always the same in every generation, or as
          severe.

Mary re-buttons her blouse.

MARY: We'll have to keep this confidential.

DOCTOR: It's nothing to be ashamed of...

MARY: I don't want anyone to know!

DOCTOR: Well you'll certainly have to tell your
          husband.

MARY: Especially not him.

DOCTOR: You're going to keep this from Richard?!

MARY: Yes.

DOCTOR: I don't think...

MARY: Dick works for the same company I do. I can't
           risk anyone at North American finding out.

DOCTOR: You're going to need support at home, Mary.
           This will be too big of a burden to handle on
           your own.

Preparing to leave, Mary puts on her jacket.

MARY: Watch me.

DOCTOR: You won't be able to hide this forever.
           Over the next few years the symptoms will
           become more acute.

MARY: I have a top secret clearance. I can't work
           at my job without it. How long do you think
           I'll be able to keep that clearance if someone
           finds out I have a mental problem?

DOCTOR: I wouldn't call it a mental problem...

MARY: That's what the government will call it!

She grabs her purse. The doctor tears off a
prescription and holds it out. Mary refuses it.

MARY: No medication. Dick might find it.

DOCTOR: I think you're making a mistake.

MARY: Let me make one thing very clear: I never
           make mistakes.

She enters the main stage as LIGHTS FADE OUT on the
doctor.

SCENE 11.

LIGHTS COME UP on the Research Department.

The names of seven rocket fuels are written on the chalkboard. They are (in order): hydrogen, hydrazine, kerosene, gasoline, diethylamine, ethylenediamine, and unsymetricaldimethyl hydrazine.

Bill Weber enters with some papers, and intercepts her.

BILL: I finished checking your calculations on the amines. As usual, your work is flawless – not a single mistake.

She takes the papers, and he walks away to exit. When they are on opposite sides of the stage Bill stops – turns around.

BILL: Oh – uh – there was one small item.

MARY: What.

BILL: The specific impulse of DETA and LOX.

MARY: Two thirty-six.

BILL: Right – that's what we use. On the hill. But the Redstone will be taking off at the Cape – at sea level. So the initial ISP will be...

MARY: (realizes her error) Two thirty-four.

BILL: A little thing. I took care of it. Good night.

MARY: A little thing! You call using the wrong ISP value little!?

BILL: It's not a big deal...

MARY: Bill: we're trying to squeeze every ounce of force out the Redstone propulsion system we

can! Two seconds of ISP isn't "little" –
especially when it means the calculated
takeoff thrust would be reduced. It's a
mistake - in the wrong direction. It's huge –
and you know it!

BILL: I know you're under a lot of pressure...

MARY: Don't be condescending! I hate it when
          the men around here do that!

Bill decides it's time to be honest and blunt.

BILL: Okay, fine. It was a huge mistake. Ya
          screwed up. I covered your ass. No one will
          know.

MARY: I will know!

Bill moves to exit. He is stopped by –

MARY: Bill. (breathe) Thank you.

Bill exits.

Mary looks at the papers Bill gave her, then sets them
on her desk. She sets her purse down.

Alone now, Mary sits at her desk and puts her head in
her hands.

She looks up – grabs a pack of cigarettes – and begins
to repetitively tap-tap-tap them into her hand
(obsessive/compulsive). She suddenly catches herself –
throws down the cigarettes – stands up – and stares at
her hands – opening and closing her fingers several
times.

She grabs a piece of chalk – then adds an eighth
chemical to the board: diethylenetriamine.

MARY: Properties and characteristics.
          Characteristics, and properties.

She meanders around the office, taking occasional looks
at the chemical formulas. She addresses the audience.

MARY: Properties and characteristics.

As she walks around the desks thinking over the
problem, Paul Morsky - carrying a card table - enters
with Michael Sherman and Mrs. Walker.

PAUL: Mary - we have a spirited game of bridge for
           you.

The three of them set up the table, and add four
chairs.

WALKER: Mary - we miss you so much.

During the following scene the three figures from
Mary's past pester and distract her as they play
bridge.

They may adlib additional lines as needed. As part of
the distraction the three of them keep changing
positions at the table - perhaps even peering over her
shoulder as she works.

MARY: There are ten properties and characteristics
           our new propellant must have. One: it has to
           be commercially available. It's no good to us
           if we can't buy the stuff somewhere.
           Two: We have to know its physical data.
           We can't work with a chemical we know nothing
           about - that's obvious.

PAUL: That's obvious!

MARY: Three: Our new fuel must have a low vapor
           pressure; it has to be a liquid at room
           temperature. The Redstone engine is not
           designed for a cryogenic fuel - the valves
           would freeze right up.

MICHAEL: Freeze right up - like a North Dakota prairie
           winter.

Mary tries not to be distracted.

MARY: Then there's the mixture ratio. The engine is
           designed to use about one pound of oxidizer
           for every three-quarter pounds of fuel. Our

HS 62

new propellant must optimize at something close to that or we won't get the maximum energy from combustion.

PAUL: Mary — it's your turn.

Mary pauses — goes to the table — picks up her cards.

MARY: What's trump?

MICHAEL: Hearts.

Standing, she plays a card — then places the remaining twelve cards back on the table. Play continues.

MARY: Five: It has to be somewhat stable. We don't want to blow up any of the technicians while they're loading it into the rocket, do we?

WALKER: Oh — that would be unpleasant.

PAUL: It sure would.

MARY: Six: Controllable toxicity. We don't want to poison those techies, either.

Mary plays another card.

PAUL: You're trumping me?

MARY: Seven: It has to have a high heat of combustion. The hotter the flame, the more energy we get out of it.

MICHAEL: Did you clean the creamer, Mary? Ya gotta clean the creamer!

MARY: Yes, daddy. You know I always clean it.

Another card — another hand.

MARY: Eight: Since the Redstone pumps the fuel around the engine to cool it off — just like a car's radiator pumps water — our new fuel must have good heat transfer characteristics. Otherwise the engine could burn up in seconds.

WALKER: We sure miss you, Mary. Everybody in Ray
wants you to come home.

MARY: Nine: It should have a low molecular weight.

Another card – another hand.

MARY: Ten: Our new propellant should have a high
Hydrogen/Carbon ratio, again for the purpose
of maximizing our combustion.
Let's see - is there anything I've forgotten?

WALKER: Did you mention density?

MARY: Oh yes – density. Density.

Mary seems lost in thought for a few moments. There's
something about density...

Another card – another hand.

MARY: Density, yes. Our fuel should have a high
density; the more molecules we can cram into
the fuel tank the better.
So that's it - just eleven little things.
Eleven properties and characteristics that
our fuel must have that make our search
essentially...impossible. The propellant we
need God has not yet created.

PAUL: God has not yet created it.

WALKER: My goodness; that is a dilemma.

She contemplates the chemicals on the chalkboard.

MARY: Let me illustrate the problem. There are
eight possible fuels that could give us the
required specific impulse.
Pure Hydrogen would be the best; high energy
and no other atoms to get in the way. But to
get a high density in the propellant tank you
have to use it in liquid form. But as we all
know, liquid hydrogen is not available in
large quantities. Either way, it's cryogenic;
so not compatible with the Redstone hardware.

She erases Hydrogen.

PAUL: You gonna talk or play cards?

She plays a card.

MARY: We could use hydrazine - commercially
           available; good. High heat of formation -
           very good. But it's extremely toxic and
           highly unstable when heated. Rockets with
           regeneratively cooled engines - like the
           Redstone - blow up whenever they try to use
           this stuff. Hydrazine is out.

She erases hydrazine.

WALKER: There we go with the "blow up" stuff again.

MARY: Kerosene. High energy - superior to alcohol,
           but it has a much different mixture ratio with
           LOX. So it's incompatible with the Redstone
           hardware.

Mary plays a card - then erases Kerosene.

MARY: Same problem with gasoline.

She erases the formula for gasoline.

MICHAEL: Gotta milk the cows, Mary. No time for
           foolishness like school.

MARY: Diethylamine has some good properties, but it
           has a low boiling point, which makes it a poor
           engine coolant.

She erases diethylamine.

PAUL: Don't worry about college, Mary. Your country
           needs you now.

She goes to play her next card - and sees something
Paul has done she doesn't like.

MARY: You led with the queen of hearts? I have the
           king. You used up my best trump card.

She tosses the card onto the table. Paul points to
Mrs. Walker.

PAUL: You told me to play that card!

WALKER: Oh sure - blame the school teacher.

MARY: ethylenediamine: C2H8N2 - so it has a high
          hydrogen/carbon ratio. But it also has a high
          vapor pressure. It's like propane; to keep it
          in a liquid state you have to keep it under
          pressure. The fuel tank would be too heavy.

She plays another card, then erases ethylenediamine.

WALKER: Think of the children. The school needs you,
          Mary.

MICHAEL: Forget about the schoolhouse. It's the
          farmhouse that's important.

MARY: That brings us to unsymmetricaldimethyl
          hydrazine. UDMH for short. This stuff was
          invented by the Soviets about eight years ago.
          Excellent performance...

Throws down another card.

MARY: ...characteristics. I'd love to use it, but
          its density is too low - we wouldn't be able
          to fit as much as we need into the Redstone
          fuel tank.

She erases unsymmetricaldimethyl hydrazine.

MARY: Last but not least - diethylenetriamine.
          Also referred to as "DETA."
                (pronounced DEE-tah)
          Not as powerful as UDMH and some of the
          others, but at point-nine-six grams per cubic
          centimeter its density is just terrific.

WALKER: (gloating--to Paul, Michael) I told you
          density was important.

MARY: Unfortunately the mixture ratio with LOX and
          DETA is off spec. Again - not compatible with

the Redstone hardware.

She tosses in the last card of her hand.

MARY: Which leaves us with...

One or more of the three card players get up and start
following her.

Mary erases diethylenetriamine. The chalkboard is now
blank.

MARY: ...absolutely nothing.

WALKER: The children need you, Mary.

MICHAEL: The farm needs you.

PAUL: Your country needs you.

MARY: Quiet! I'm trying to think!

MICHAEL: (to Paul) Did you know the specific impulse of
           DETA and LOX at sea level is two hundred and
           thirty-four seconds?

PAUL: Oh yeah.

WALKER: Well of course - everybody knows that.

MARY: Shut up! SHUT UP!

LIGHTS FADE TO BLACK

END SCENE 11.

SCENE 12.

LIGHTS UP on the office of Tom Meyers. Early morning -
the engineers are not yet at their desks.

Tom is working at his desk. A few beats, then Don -
nervous and breathless - runs in.

DON: He's here! He's in the building!

TOM: Who?

DON: Mister Big. Mr. Mucky-Muck himself.

TOM: Phil Ellsworth? So what.

DON: No! Not that butt-scratcher! I mean the _real_
          Mr. Big - The Legend - Dutch Kindelberger!

TOM: Nice try, Jenkins. Kindelberger hasn't
          visited this plant in ten years.

DON: Hand to god - and he's heading this way.

Tom is not sure if he should believe him.

DON: You better get that desk cleaned up, pardner.

Don quickly exits. Tom reflects a couple of beats,
then decides Don could be telling the truth, and rushes
to clean up his desk before -

Dutch Kindelberger enters. He is in his late sixties,
ramrod straight, with a tailored suit and white shirt.
A cloud of absolute authority hovers over his head.

DUTCH: Hello, Tom.

TOM: Mr. Kindelberger.

DUTCH: Mind if I come in?

TOM: It's your company, sir; you can go wherever
          you want.

DUTCH: I know. But courtesy, Tom - never forget it.
          Courtesy goes a long way.

Dutch makes himself at home.

DUTCH: How have you been, Tom?

TOM: Great.

DUTCH: The wife - kids?

TOM: We're all doing just fine.

DUTCH: Wonderful. Having a family is a wonderful
            thing, isn't it?

TOM: Yes, sir; it is.

DUTCH: And being able to support them; that's a
            wonderful thing, too.

TOM: Absolutely.

DUTCH: Hope you don't mind me barging in like this,
            Tom, but I keep getting calls from all sorts
            of people - politicians, army generals, mucky-
            mucks up the wahzoo. Wernher von Braun calls
            me almost every day.

He is lost in thought for a beat or two.

DUTCH: Wernher von Braun. Ya know, in World War II
            our company built thousands of planes and
            bombers designed to kill the sonovabitch. Now
            we work for him. Half the contracts we get
            have his signature at the bottom.

TOM: Strange how life works out.

DUTCH: There's a lot at stake here, Tom. I've given
            my heart and soul, my life, and every penny I
            have to build this company. We have a great
            reputation with our customers. Now that
            reputation is on the line.

TOM: I know that, sir. And you know we've always
            put out a quality product.

DUTCH: Of course. Still - there's this girl. What's
            her name - the one you got on the propellant
            contract.

TOM: Mary Morgan.

DUTCH: Right. How's she comin' along?

TOM: Real good. Making steady progress.

HS 69

DUTCH: Don't bullshit me, Tom.

TOM: She's...struggling; but remember – not even
            von Braun's best engineers could solve this
            problem.

DUTCH: Hm. You know I trust you.

TOM: I hope so.

DUTCH: I know you wouldn't do anything to jeopardize
            my company – or your job. Would you.

TOM: No sir. I'm very happy working at North
            American.

LIGHTS FADE UP on engineer's desks.
Mary quietly enters and sits at her desk.

DUTCH: We're in an undeclared space race here, Tom.
            Russia puts a satellite into orbit one day,
            and the next day America is some snot-nosed
            pooper grabbing his mommy's dress and crying
            to have his diaper changed.
            I always said, "First country to put a
            satellite into orbit will be on top of the
            world." Ya can't buy that kind of prestige
            with money, Tom. You buy it with grit. You
            buy it with risk. You buy it with balls.
            "The whole world," I said. "The whole world
            is going to look to that country as a leader."
            And the company that puts them there – they're
            gonna get a helluva lot of contracts.
            We don't want the world doing business with
            the Russians, do we, Tom? Hell – they're all
            a bunch 'o communists – whatta they care about
            business anyway?

TOM: With all due respect, sir, America could have
            been in orbit a year ago if we had an engineer
            in the White House instead of a politician.

DUTCH: Maybe you should run next time.
                 (a beat or two)
            Two hundred years from now some star-struck
            fifth grader will be sitting in physical
            science class, and his teacher will tell him

how Russia was the first country to place a
satellite into orbit. It's an accomplishment
that will live forever. No one will ever be
able to take it from them.

TOM: That's true, sir.

DUTCH: This girl - Morgan. Any chance things would
          go a little faster if we put somebody else on
          it?

TOM: I honestly don't think so. And even if I
          found someone better, I wouldn't do it.

DUTCH: Why the hell not!?

TOM: Because, sir; courtesy goes a long way.

Tom's use of Dutch's opening advice catches the man off
guard.

DUTCH: Yes. Yes it does.
              (all's quiet)
          Well - I guess that's that.

Dutch retrieves a golf club from Tom's golf bag and
holds it up in what appears for a moment to be a
threatening gesture.

He then takes a drive position and takes a practice
swing.

DUTCH: You going to be at the Downey Golf Tournament
          this year?

TOM: Wouldn't miss it, sir.

DUTCH: Good.

Dutch puts the golf club on Tom's desk.

DUTCH: Ya know, Tom - you outta try to keep your
          desk a little cleaner.

TOM: I will, sir.

Dutch moves to the door.

HS 71

DUTCH: Tell Gloria I said "hi."

Dutch exits.   Tom lets out a long exhale.

Mary is working alone at her desk.   She sees Dutch
leave Tom's office.

She works quietly for a few moments, then Tom exits his
office and approaches.

TOM: Mary?

MARY: Yes, Tom.

TOM: Could I have a minute?

He has trouble starting — something is on his mind.
She reads it.

MARY: You want to know how the Redstone project
          is going.

TOM: Yes I do.

MARY: And you want to know before our usual Monday
          morning briefing.

TOM: Yes.

MARY: We've eliminated the Amine compounds. We're
          going to take another look at a couple of the
          hydrocarbons...

TOM: You know what I mean.

MARY: I am going solve this, Tom.

TOM: How much longer do you think it will take?

MARY: I don't know, maybe...

TOM: (loses his cool) How much longer!?

His anger quiets her and she's not certain how to
respond.

TOM: This chalkboard. This desk. These books.
            This slide rule. Simple, innocuous objects.
            Nothing glamorous or special about them.
            But they're the same tools the Russians
            have — and they managed to get into orbit
            using them. We promised the Army we'd have
            this done two months ago — and you're no
            closer than you were when you started.

MARY: That's not true — we're a lot closer. It's a
            process of elimination. We keep eliminating
            propellants that won't work till we find one
            that does.

TOM: How many have you eliminated?

MARY: Well...all of them, but...

TOM: So that's it? We're all done? Should I tell
            the Army we failed? Do you have any idea how
            much of our paychecks depends on government
            contracts? Don't answer — I'll tell you: all
            of it! Everything we work on in this
            department gets its funding from the Feds.
            You know what happens when they award a
            contract and you fail? You know what they
            do!? Don't answer — I'll tell you; they send
            the next twenty contracts to our competitors!
            And we hand out pink slips!

MARY: What do you want, Tom?

TOM: I want to know when you're going to finish.

MARY: What do you want, Tom?

TOM: I want to know when you're going to finish!

MARY: What do you want, Tom!?

TOM: (shouting) I want a deadline!

MARY: I can't give you one!

TOM: The rocket is sitting on the pad at the Cape.
            It's fuel tank is empty. A thousand engineers
            and technicians are waiting for us to tell

them what to fill it with. Our country's
entire space program is on hold - and it's
holding for you, Mary.

MARY: I saw you had a visit from the Dutchman.

TOM: Oh yeah.

MARY: So, what; shit runs downhill?

TOM: They haven't changed the laws of gravity.

MARY: Tell them I'll have it done by next Friday.

TOM: Really?

MARY: Really.

TOM: Next Friday's good. Next Friday's very good.
           And I can tell the customer?

MARY: Sure.

TOM: Good. Thank you. Thank you!

Tom exits, noticeably happier.

Mary pages through the reference book for a bit, then
slams the book down hard on her desk.

She grabs her hair, breathing heavily. She releases
it. She picks up a pack of cigarettes and starts
nervously tap-tapping it into the palm of her hand.
She paces the room.

A few beats - then she catches herself, and puts the
cigarette pack back down - then leans against her
desk. A beat or two to calm herself down.

LIGHTS OFF A FEW BEATS - THEN LIGHTS UP

Mary is sitting at a different part of the office
studying another reference book.   Tom enters.

TOM: It's Friday. Do you have it?

MARY: Looking good for next week, sir.

HS 74

TOM: Another week! You said...

She holds up an index finger to shush him. He turns
and exits.

LIGHTS OFF A FEW BEATS - THEN LIGHTS UP

Mary is working at another desk. Tom enters.

TOM: It's Friday.

MARY: One more week.

Tom reluctantly exits.

LIGHTS OFF A FEW BEATS - THEN LIGHTS UP

Mary is at yet another desk, working with a slide
rule. Tom enters.

TOM: It's Fri...

MARY: Another week.

TOM: This is it, Mary. It's over. You don't
          have it by next Friday, I'm giving the
          project to Irving.

Tom exits.

LIGHTS OFF A FEW BEATS - THEN LIGHTS UP

A disheveled Mary is lying on her back on one of the
desks working a slide rule and groggily singing "Fly
Me To The Moon."

Bill Weber enters, carrying a mug of coffee.

BILL: Good morning. (sits at his desk)
          You haven't been here all night, have you?

MARY: We need von Braun to increase the size of
          the fuel tank.

BILL: Not gonna happen.

The phone RINGS on Mary's desk. She walks over and picks up.

MARY: Research. (a beat) Yes, this is her.

A few beats, then Mary quietly sets down the receiver.

BILL: Something wrong? (no response) What is it?

MARY: My father just passed away.

BILL: Oh Mary – I'm so sorry. Is there anything
          I can do?

She stands up and faces him.

MARY: Yes. You can figure out how to put
          Unsymetrical Dimethyl Hydrazine into the
          Redstone, and make it work.

LIGHTS FADE BLACK as Mary exits.

END SCENE 12.

SCENE 13.

A confessional in a Catholic Church in Ray, North Dakota.

Father Mackey, dressed in his priestly blacks and white collar, sits on his side of the confessional.

Mary enters – wearing a black lace veil. She faces the audience and genuflects once – making the sign of the cross, then enters the confessional.

MARY: Bless me, Father, for I have sinned. It has
          been one month since my last confession.

FATHER: Mary? Is that you?

MARY: Yes, Father.

FATHER: It's good to have you back in Ray. Unburden
          your soul, my daughter.

MARY: Father, I have committed the sin of anger.
          And I have lied.

FATHER: Lied? That doesn't sound like you. Who did
          you lie to?

MARY: The Federal Government of the United States of
          America.

FATHER: Hoo boy.

MARY: Although, I'm not exactly certain if it was a
          lie, per se.

FATHER: What did you tell them?

MARY: This contract I'm working on; I told them I
          would have it ready four weeks ago.

FATHER: Did you know it would not be ready?

MARY: Yes.

FATHER: Well yes, that was a lie.

MARY: And I've been angry with my boss – and some of
          my co-workers – and my husband.

FATHER: Husband?

MARY: Oh, I forgot to tell you. I got married.

FATHER: Married! A good Catholic boy, I assume.

MARY: No, Father.

FATHER: Oh. Any chance he might convert?

MARY: I don't think so.

FATHER: That's disappointing. You know how I feel –
          people should always marry within their own
          faiths. It just works out better.

MARY: I know, Father. But I was so conflicted; he's
          handsome, and witty, and a brilliant engineer.

                    (a beat)
                Although, he's kind of a mediocre bridge
                player. But I'm teaching him.

FATHER: Mary — do you really believe any of those
                reasons are substantial enough to lead one to
                a life-long commitment? What else can you
                tell me about him?

Mary really wants the priest to be impressed with
Richard.

MARY: Well...he played college football in the
                Rose Bowl.

FATHER: Did he!

MARY: More than once.

FATHER: More than once! Goodness - now I'm conflicted.
                    (a beat)
                You said you were having a problem with anger.
                What is the source of this anger?

MARY: I'm under a lot of pressure at work. The
                stress is unbelievable.

FATHER: Tell me about it.

MARY: You know, Father, my work is all top secret.
                I would probably be breaking the law by
                discussing it with you.

FATHER: The power of the confessional will protect
                you. Now tell me how it's going.

MARY: So far it's a complete failure. I'm a
                complete failure. I thought I could do it,
                but the computations and permutations -
                there's just too many of them. It's all too
                complex. They're asking me to do the
                impossible.

FATHER: Did not our Father in Heaven ask the same
                thing of our Lord Jesus?

MARY: Yes, Father.

FATHER: Christ came to Earth to show us the way to
        live. Everyone has moments in their life when
        they are faced with having to accomplish the
        impossible. This is your moment, Mary. You
        need to remember and ponder the example of our
        Lord and Savior.

MARY: Yes, Father.

FATHER: What is it about your work that seems so
        impossible? Is it that fuel problem you were
        telling me about?

MARY: Dr. von Braun is already using what I consider
        the perfect propellant combination for the
        Redstone: LOX and alcohol...

FATHER: LOX?

MARY: That's short for liquid oxygen.

FATHER: Ah.

MARY: I'd like to substitute the alcohol with
        unsymetricaldimethyl hydrazine, but the
        density is way too low.

FATHER: Take it from an old Irishman - there's no
        substitute for alcohol.

MARY: The problem, Father, is that God has not yet
        created the fuel we need.

FATHER: Well then, maybe you should create it. God
        never built an airplane - that didn't stop the
        Wright Brothers.

MARY: Yes, Father.

A long sigh from the priest. How does he give advice
on this subject?

FATHER: Every evening after dinner I have a small
        glass of brandy. It helps me unwind.
        Recently I started adding a pinch of Tabasco
        sauce to give it an extra kick. Just a small

amount - but it makes a big improvement. Maybe
that's what you need. Instead of replacing
this fuel, you should just add something to
it. You know - give it an extra kick.

Mary stands up and steps out of the confessional -
standing just outside of it, lost in thought. She has
a "Eureka" moment - and exits on a run.

The priest is unaware she has gone.

FATHER: Of course, I'm really not qualified to give
         advice on this subject. Anyway - those are my
         thoughts on the matter. For your penance say
         three Hail Marys and two Our Fathers.

He begins to give her a Latin blessing...

FATHER: Dominus noster Jesus Christus te absolvat; et
         ego auctoritate ipsius...

...then suspects she's not there.

FATHER: Mary? (steps out) Mary?

LIGHTS FADE TO BLACK.

END SCENE 13.

SCENE 14.

LIGHTS UP on the Research Office. Irving, Joe and Bill
are at their desks.   There is a dark, somber mood. No
one speaks - it's as if they are at a funeral. In Tom's
office, Don and Tom are having a quiet meeting.

Joe aimlessly tosses crumpled balls of paper into a
waste basket ten feet away.

Mary enters on a run - breathless.

MARY: UDMH and DETA - are they miscible?

No one pays attention.

HS 80

She pages frantically through a reference book.

MARY: Did you hear me? Are they miscible?
              (more paging)
          Where's Richard?

BILL: He's on the hill - doing a test on the new
          engine. Is what miscible?

MARY: UDMH and DETA.

IRVING: Yes - they're miscible.

BILL: What does that matter?

MARY: I've been going about this all wrong. All this
          time I thought it was a process of elimination.
          It's not elimination; it's creation. We're
          going to create a brand new fuel.

She paces.

MARY: We discarded unsymetricaldimethyl hydrazine.
          Why. Because of the density problem. But who
          says the fuel has to be all UDMH? What if we
          remove a small portion, and replace it with a
          denser fuel, like diethylenetriamine. You'll
          get some of the benefits of both: a higher
          specific impulse, and a better density. It's
          like a rocket cocktail - like adding Tabasco
          sauce to brandy. The only question now is -
          how much UDMH do we remove, and how much DETA
          do we add. Ten percent? Fifteen percent? It
          doesn't matter - because now it's just a matter
          of number crunching - to find the ideal
          mixture. This is it! This is the answer!

BILL: (insincere) That's great, Mary.

IRVING: Yeah. We're very proud of you.

For the first time Mary notices the somber mood.

MARY: You don't seem very excited. Don't you get it
          - this is the answer. We're going into space!
              (a beat)
          You guys look like you're at a funeral.

JOE: You didn't hear the news?

MARY: What's happened?

IRVING: This morning the Russians placed a dog into
          orbit.

MARY: A dog? You mean a live dog?

BILL: Yep. It's up there. Barking and skipping and
          shitting through space.

MARY: Why put a dog...

JOE: They're experimenting with life support
          systems.

IRVING: First you put up a satellite...

BILL: Then an animal...

MARY: ...then a human.

JOE: They're going to have a man in orbit before we
          have so much as a nut or bolt.

Mary collapses into her desk chair, joining the somber
mood.

IRVING: If a Russian dog barks at an American dog,
          do they understand each other?

Mary stands up and retrieves the card table. She sets
it up, then rolls her desk chair next to it. She grabs
a deck of cards, sits down, and starts dealing four
hands.

JOE: Mary - it's not lunch time.

She keeps dealing. Bill rolls his chair over and joins
her. A beat or two, then Joe rolls over.

Don enters from Tom Meyers' office.

DON: Mary. Mister Meyers wants to see you in his
          office.

MARY: Not now.

DON: He said it was urgent.

MARY: Tell him I'm playing bridge and I cannot be
           disturbed.

DON: I — I can't tell him that.

BILL: (shouting to Tom) She's playing bridge and she
           cannot be disturbed!

TOM: Okay!

Tom CLUNKS a large glass onto his desk, then CLUNKS
large bottle of scotch onto it as well.

Don wheels his chair over, becoming the fourth player.

DON: This department is getting stranger by the
           day.

The cards are dealt and the players begin the bidding.

JOE: Two hearts.

BILL: So we get the higher performance by using
           UDMH...

DON: Two spades.

JOE: ...and we solve the density problem by mixing
           in some DETA.

BILL: Four hearts.

DON: That's a pretty good idea.

Irving stands in kibitz position.

IRVING: I would have thought of that eventually.

MARY: Four spades.

The other three players "pass." The first hand begins.

IRVING: Play the diamond.

LIGHTS FADE TO BLACK as the bridge game continues.

END SCENE 14.

SCENE 15.

LIGHTS UP on the Research Office, two weeks later. All
the engineers are present. A party. Drinks. A
punchbowl.   Hors d'oeuvres.

Richard and Joe are at their desks holding small
foreign language help books and conversing in basic
Japanese.

The phonetic pronunciation is listed for each sentence.

JOE: Mor-u'go-san, oh gen-key dess' ka?

RICHARD: High, gen-key dess. Foo-ree-doh-mahn-sahn
          wah?

JOE: High, oh kah-geh sama de.

RICHARD: Oh the-ah-rye wah doh-koh dess' kah?

Mary enters carrying a coffee cup.

MARY: Hi, Honey.

She heads to her desk.

JOE: Reh-stoh-rahn no u-rah ni ah-ree-mahss'

RICHARD: Doh-moh a-ree-gah-toh go-zye-mahss'.
          Ee-mah ee-kee-mahss'.

MARY: How's the language lesson going?

RICHARD: It's going foo-tsoo yo-ree oi-shee dess'.

Both men laugh.

MARY: You're not funny.

Joe points at Richard.

JOE: Oh-kash'-koo-nye dess'! Oh-kash'-koo-nye dess'!

Both men laugh even harder. Mary rolls her eyes. Richard stands up and grabs a small suitcase.

RICHARD: Wish me luck.

He and Mary peck a kiss.

MARY: Good luck. Got your passport?

He pats his coat pocket, then heads for the door.

JOE: Mor-u'go-sahn, saw-yo-nah-rah!

RICHARD: Foo-ree-doh-mahn-sahn, saw-yo-nah-rah!

JOE: Saw-yo-nah-rah!

RICHARD: Saw-yo-nah-rah!

MARY, DON, IRVING, JOE: Go!

An entering Tom passes the exiting Richard.

TOM: Sayo — whatever.

Tom joins the party. He approaches Mary.

TOM: Just got back from talking to the mucky-mucks upstairs. They've approved your research.

Everyone applauds.

TOM: Colonel Wilkins will be by any moment to pick it up.

Tom removes a letter from his coat pocket.

TOM: I received a letter this morning that I'd like to read.
                (unfolds the paper)
         Dear Tom — please offer my congratulations to Mrs. Morgan for her fine work with the

HS 85

Redstone propellant project. With superior
engineers like her I am confident our company
will have a bright and prosperous future.
Best Wishes, Dutch Kindelberger.

TOM: You probably noticed he used the word
          "engineer." I suppose if calling Mary
          an engineer is okay with Mr. Kindelberger,
          then it's okay with me.

Tom shakes Mary's hand.

TOM: Congratulations.

Her fellow engineers applaud.   Colonel Wilkins enters.

TOM: Colonel - you're just in time. I was about to
          discuss naming our new propellant. Since the
          right to name any new invention traditionally
          goes to its inventor - Mary; what do you think?

MARY: Well, I think we should call it bagel.

TOM: Bagel?

MARY: Yes - that way we can say the Redstone's
          propellant combination is LOX and bagel.

Her fellow engineers laugh, but not Tom or the Colonel.

TOM: I don't think the boys upstairs are gonna go
          for that.

MARY: Then let them name it. I've got work to do.

COLONEL: If it's all the same to you, the Army has
          already chosen a name. We want to call it
          hydyne.

MARY: Hydyne. Not as funny as bagel.

COLONEL: The Army believes this is going to be the
          propellant of the future.

He takes a step toward Mary and they shake hands.

HS 86

COLONEL: The Army thanks you for your service.
Mr. Meyers - if you have a moment there's
another contract I'd like to discuss.

The Colonel follows Tom to his office. At that moment
DAVID SHELBY - wearing a "VISITOR" badge and carrying a
small notebook - enters.

DON: Who's this?

DAVID: I'm looking for Mrs. Morgan.

MARY: That's me.

DAVID: Hi - I'm David Shelby from Life Magazine. I'd
like to discuss doing a feature article on you.

MARY: No. Get out.

DAVID: Get out? I said Life Magazine.

IRVING: You heard her - get out.

BILL: Mary - maybe you should talk to him.

DAVID: Mrs. Morgan - there's an entire world out there
completely oblivious to what you have
accomplished here. Life Magazine is going to
be your partner to fame and fortune. We're
gonna make Mary Sherman Morgan a household
word.

MARY: That's three words.

DAVID: I've already written most of the article;
I just need to fill in a few details.

MARY: I want you to leave.

BILL: Mary...

DAVID: Leave? You can't be serious.

MARY: You think everybody wants to be famous?
Everybody wants to be a celebrity? Some of us
are perfectly happy living quiet, anonymous
lives. Does that surprise you?

DAVID: Yes!

MARY: If you print anything with my name in it, I'll
          sue you — and your publisher.

JOE: I think it's time for you to go.

The engineers, except Bill and Mary, intrude on the
reporter's space.

DAVID: This is insane! You've made history here —
          people are going to want to know about it.

The engineers, except for Bill, steer David toward the
exit. He struggles to remain.

DAVID: STOOOOOP!

Everyone stops. He turns and runs back to Mary's desk.

DAVID: Mrs. Morgan - the Media is mankind's last great
          hope for immortality! If you push me away,
          whatever legacy you might have had will
          gradually be extinguished — like a light fading
          out in a darkened room!

The male engineers grab him bodily and toss him out of
the office. Off stage we hear the sound of large heavy
objects crashing.

Tom comes out of his office to investigate. Irving stops
him.

IRVING: Everything's under control.

Tom's puzzled, but he returns to his office.

DON: I've done my work for the day.

Don, Joe and Irving put on their suit coats. Don and Joe
grab their briefcases and, with Irving, say their good-
byes and move to exit. Don notices Irving has forgotten
his briefcase - again.

JOE: Hey — Mister Forgetfull.

Irving remembers – returns for the briefcase – and the
three of them exit. Mary and Bill are left alone.

MARY: Life Magazine. I wonder how on earth they got
           my name.

She grabs her purse and moves to exit. She slows, stops,
realizes the truth, then returns.

MARY: You called them. Didn't you. (beat) Bill?

BILL: I just felt it was time for you to get a little
           notoriety for your work.

MARY: I don't want notoriety – I'm not some publicity
           hound like von Braun. All I want out of life is
           a husband that loves me, a job that challenges
           me, and bridge partners that are skilled enough
           to beat me every once in a while.

BILL: Yeah, well, I guess we're still working on
           number three.

She pulls her car keys from her purse.

MARY: Please don't do that again.

Mary exits – leaving a dejected Bill alone.

LIGHTS FADE TO BLACK.

In complete darkness, an audio recording is played of the
last twenty seconds prior, and first thirty seconds
after, the launch of Explorer I. An audio copy of this
recording should be found at:
http://www.youtube.com/watch?v=JgguaLjyna8&NR=1

END SCENE 15.

SCENE 16.

A few beats – then a small campfire breaks through the
darkness. A sooty metal coffee pot sits close, its
contents warming.

The sound of WIND – not too harsh.

The sounds of BRANCHES BEING BROKEN and AXED IN HALF off stage is heard. A few beats of silence, then the sounds are repeated.

Richard, wearing a warm jacket, enters the firelight carrying an armload of branches. He sets down the firewood and places one branch on the fire.

There is a log for sitting, and he takes a seat. He grabs a metal cup and pours some coffee into it. He stokes the fire, then puts on another branch. He takes his time – he's on vacation.

Then Mary - also dressed warmly – enters, carrying more firewood. She drops the wood by the fire.

Mary leans over the fire and warms her hands.

RICHARD: Great day today.

MARY: It was wonderful.

He hands her the coffee cup, then fills his own.

RICHARD: Think we'll see any bears out here?

MARY: I sure hope so.

Mary sits next to him on the log and they snuggle close.

MARY: Hope you didn't mind having to wait for the honeymoon.

RICHARD: Nah.

MARY: Really?

He holds her tight, she sips some coffee, and they watch the fire.

Their peace is interrupted by the sound of APPROACHING FOOTSTEPS. They look up to see a woman approaching – a Yellowstone Ranger.

RANGER: Howdy, folks!

MARY: Good evening.

RANGER: Welcome to Yellowstone – the world's first
           national park. Saw your fire and thought I'd
           come on over. Don't get many campers this time
           of year.
                (looks around)
           You practically have the place to yourselves.

MARY: Fine with us.

RANGER: Been having a good time?

RICHARD: You bet. We drove up to see Old Faithful
           this afternoon.

RANGER: Ah, Old Faithful. Yessiree. You take a lot of
           pictures?

RICHARD: Oh yeah.

He holds up his camera with the prototype zoom lens.

RANGER: What's that on your camera?

RICHARD: It's a zoom lens – a new design that's a lot
           smaller and lighter.

RANGER: May I see it?

Richard hands her the camera.

RANGER: I'm kind of an amateur photographer.
                (examines the lens)
           Boy-howdy, that is a lot lighter. Big
           improvement.

MARY: Would you take our picture?

RANGER: Sure!

She steps back and focuses the lens.

RANGER: Y'all say "grizzly."

RICHARD AND MARY: Grizzly.

HS 91

She takes the picture, then hands the camera back.

RANGER: Seen any other geysers since you've been here?

Richard and Mary kiss — and continue enjoying each
other's company during the geyser dialogue.

MARY: Several.

RANGER: You all know how a geyser works?

MARY: Surface water seeps through rocks until it
          comes in contact with super-heated magma.

RICHARD: The water at the bottom of the geyser becomes
          geothermally heated, while the water at the top
          cools off.

MARY: This causes a convection effect, with the
          heated water rising to temperatures above
          the boiling point.

RICHARD: Steam rises to the top, causing a release
          of pressure...

MARY: ...forcing a combination of steam and water
          to expand and spray out the geyser's vent.

The ranger senses there's something more going on than
geyser talk, but keeps her professional composure.

RANGER: That's absolutely correct!

MARY: Well, you know — it's not rocket science.

RANGER: No, ma'am. Well — Y'all have a good evening.

The Ranger exits, and Richard takes a last sip of
coffee. Peace and quiet for a few beats.

MARY: The engineering department at the Downey plant
          is having a bridge tournament next week.
          I signed us up.

The WIND PICKS UP — louder now — and it blows the fire
out as LIGHTS FADE TO BLACK.

HS 92

<u>END SCENE 16.</u>

SOUND OF WIND continues — then FADES OUT.

<u>SCENE 17.</u>

LIGHTS FADE UP on a bar in Los Angeles. Irving is
sitting on a bar stool, nursing his drink. His
briefcase rests on the floor.

He stirs his drink — then downs the whole thing. He
speaks to an unseen bartender.

IRVING: I need a refill. Hey — buddy!

He holds up his empty glass. The bartender takes the
empty glass.

Don joins Irving on a second barstool.

DON: Looks like it's gonna rain.

The bartender brings Irving his drink.

DON: A Coke.

The bartender leaves to fetch it.

IRVING: Coke?

DON: It's the pause that refreshes.

Irving takes a sip — then begins to expound.

IRVING: Ya know — I'm not just some dimwit with a
         degree — right?

DON: You invented inhibited red fuming nitric acid.

IRVING: Bet your ass.

The bartender gives Don his drink.

IRVING: I got a big future in the aerospace business.
           A big future. Rocket to the toppet.

DON: Rocket to the toppet.

They clink glasses, then down their drinks.

Irving throws some money on the bar.

IRVING: See ya Monday.

He walks out — conspicuously leaving behind the
briefcase. The bartender takes Irving's glass.

DON: Where's your bathroom?

The bartender points. Don leaves some money on the bar,
then exits.

LIGHTS FADE TO BLACK — except ONE SPOT on the briefcase.
Then the SPOT BLACKS OUT.

END SCENE 17.

EPILOGUE

LIGHTS UP on the obit department of the LA Times.
Barbara is working. Carley enters.

CARLEY: Hi Barbara.

BARBARA: Miss Morgan.

CARLEY: I sent you the signed affidavits — did you get
         them?

BARBARA: I did.

CARLEY: So do you have everything you need? Can we get
         my grandmother's obituary published?

BARBARA: No.

CARLEY: No!? Why not!?

Barbara picks up the affidavits on her desk.

BARBARA: I told you - we need independent verification.
You are the source of all this information.
It doesn't qualify as independent.

CARLEY: An obituary is a person's last chance to tell
the world who they were and what they
accomplished. If you don't print it, my
grandmother's legacy will disappear forever.

BARBARA: I'm sorry.

Carley turns and finds Mary standing there. They hug.

CARLEY: I'm sorry, grandma.

MARY: It's not your fault.

CARLEY: I'll never forget you.

They release - and Carley backs up into the darkness.
Mary watches her go, then turns to the audience.

Her fellow engineers, wearing their suit coats, filter
in and stand near their desks, also facing the
audience.

MARY: With the launch of Explorer I, America
entered the space race. It would be the first
and only time hydyne would be used in any
American rocket. My invention, though
significant at the time, would soon fade into
obscurity.
(a beat)
After making several trips to Japan, Richard
and Joe discovered the lens companies had
replicated their design without licensing it.
To this day they have never received one penny
in royalties.

Joe Friedman turns off his desk lamp, grabs his
briefcase, and exits.

MARY: As a result of leaving behind a briefcase full
of secret documents at a bar in Los Angeles,

Irving was fired from his job at North
American Aviation. In order to get his job
back he studied law and became an attorney.
His lawsuit against North American failed, and
a year later he began work as a criminal
defense attorney. In 1972 Judge Charles Older
of the Los Angeles Municipal court assigned
Irving Kanarek to be the leading defense
attorney for Charles Manson. Vincent Bugliosi
would later refer to him as the "Toscanini of
Tedium."

Irving Kanarek turns off his desk lamp, forgets his
briefcase, and moves to exit. Mary and the other
engineers cough or clear their throats to get his
attention.

He realizes his mistake, returns for the briefcase,
then exits.

MARY: Don eventually met a nice girl, and settled
          down.

Don Jenkins turns off his desk lamp, grabs his
briefcase, and exits.

MARY: The research paper Bill and I published in
          1957 is still used by rocket and aerospace
          companies throughout the world.

Bill Weber turns off his desk lamp, grabs his
briefcase, and exits.

MARY: Richard spent forty years working in the
          Aerospace industry, participating in every
          Major project; Redstone, Mercury, Gemini,
          Apollo and the Space Shuttle. He never again
          Played football in the Rose Bowl.

Richard Morgan turns off his desk lamp, grabs his
briefcase, and exits.

MARY: As for myself, after fifteen years in the
          aerospace business I retired to become a full
          time wife and mother. Richard and I had four
          children, and we lived happily ever after.
               (a beat)

In August, 2004, after fighting numerous
illnesses related to thirty years of smoking,
I passed away. My family's attempts to
convince the Los Angeles Times to publish my
obituary were ultimately unsuccessful. My son
was so angered by this, he decided to write
this play. As of today, ___*(today's date)*___ ,
the Los Angeles Times still refuses to publish
my obituary.
My name is Mary Sherman Morgan. I am an
engineer.

LIGHTS FADE TO BLACK

<u>END</u>

GEORGE D. MORGAN — AUTHOR

George Morgan began his writing career at the age of sixteen, writing and producing a musical play. He has since written dozens of works, including stage plays, musicals, music scores, screenplays and novels.

George's memoir on his mother's life was published by Prometheus Books in July, 2013.

George is a member of the Writers Guild of America. He and his wife Lisa live in McKinney, Texas.

BRIAN BROPHY — DIRECTOR

Brian Brophy is a member of the Screen Actors Guild and has appeared in many movies and TV series including SHAWSHANK REDEMPTION, ARMAGEDDON, and KISS THE GIRLS. His portrayal of Commander Maddox in STAR TREK: NEXT GENERATION made him famous as the "man who wanted to dismantle Data."

Brian is now the Director of Theatre Arts at Caltech. He and his wife Cynthia have two daughters and live in Los Angeles.

Mary Sherman Morgan at North American Aviation, 1955

Please visit www.georgedmorgan.com for information
about licensing this play, or write to:
wgawriter@aol.com

Rocket Girl: now a bestselling book. Available through amazon.com, bn.com, itunes.com, or your local bookseller.

Other plays by George D. Morgan:

Nevada Belle
Pasadena Babalon
Thunder in the Valley
Second To Die
The Trial of Goldi Locks (music and lyrics)
Capture The Sun
Closing Credits: An Evening With Frank Capra (with Loren Marsters)

HS 100

www.ingramcontent.com/pod-product-compliance
Lightning Source LLC
Chambersburg PA
CBHW070829180526
45168CB00002B/781